The Interpretation of Ionic Conductivity in Liquids

The Interpretation of Ionic Conductivity in Liquids

Stuart I. Smedley

Victoria University of Wellington
Wellington, New Zealand

PLENUM PRESS · NEW YORK AND LONDON

Library of Congress Cataloging in Publication Data

Smedley, Stuart I
  The interpretation of ionic conductivity in liquids.

  Includes index.
  1. Electrolytes—Conductivity.   2. Ionic solutions.   I. Title.
QD 561.S477   541.3'72   80-17941
ISBN 0-306-40529-6

© 1980 Plenum Press, New York
A Division of Plenum Publishing Corporation
227 West 17th Street, New York, N.Y. 10011

Printed in the United States of America

To my parents, Jack and Margaret Smedley

Preface

The phenomenon of electrical conductance in liquids is of great importance to the technologist, as well as to the theoretical scientist. A glance at *Chemical Abstracts* will reveal that electrical conductivity can be used as an analytical tool for such diverse substances as concrete and suntan lotion as well as a tool for elucidating the dynamics of molecules in simple liquids. It is a phenomenon that is relatively easily measured, which explains the great diversity of conductance studies that span a range of experimental conditions unequalled in the study of nonequilibrium phenomena. It is clearly impossible for one book, notwithstanding the ability of one author, to cope with so much information or to cover even a significant fraction of the literature on this subject. However, I believe it is possible to bring together in one monograph the mainstream ideas on the interpretation of the phenomenon in relatively simple systems. It is hoped that this book will achieve this result and will provide a concise and coherent account of the interpretation of ionic conductivity in dilute electrolyte solutions, concentrated solutions, low-temperature or glass-forming molten salts, ionic melts, molecular fluids, and fluids of geological and industrial interest. Most of these topics have been discussed in other books and review articles, but to the best of my knowledge they have not been gathered together in a single monograph.

No attempt has been made to describe the experimental apparatus used to measure electrical conductivity in liquids, and the data quoted here are regarded as reliable, unless stated otherwise. Details of experimental technique can be obtained from the original literature.

I would like to acknowledge the help of Professor J. W. Tomlinson for guidance during the formulation of this book and for subsequent discussion and critical comment. I should also like to thank Dr. R. A. Matheson, Mr. D. R. MacFarlane, Mr. J. Satherly, and Dr. J. Tallon for reading chapters of the book. My sincere thanks to Mr. E. Stevens for drawing the diagrams and to Mrs. M. Povey for her skill in deciphering my handwriting, correcting errors, typing the manuscript, and organizing my reference material.

Victoria University of Wellington   Stuart I. Smedley

Contents

3. Ionic Conductivity in Low-Temperature Molten Salts
   and Concentrated Solutions

4. Electrical Conductivity in Ionic Liquids at High Temperatures

5. Ionic Conductivity in Molecular Liquids and Partially
   Ionized Molten Salts

6. Electrical Conductivity in Liquids of Geological and
   Industrial Interest

Notation

Capital letters are listed before lowercase letters. Greek letters are listed at the end.

A . . . . Preexponential constant in equation (3.12)

$\bar{A}$ . . . . Preexponential constant in equation (3.8)

B . . . . Constant in the VTF equation (3.12)

B . . . . Jones–Dole coefficient, Chapter 1

B' . . . . Constant in the VTF equation (3.16)

C . . . . Constant, Chapter 3

$C_{p,i}$ . . . . Partial molar heat capacity of ion i

D_c . . . . Constant

D_i . . . . Diffusion coefficient of ion i

D_{12} . . . . Diffusion coefficient of molecules 1 in host molecules 2

D_α . . . . Diffusion coefficient of molecule α

E . . . . Electric field

E . . . . Energy level, Chapter 3

E_1, E_2 . . . . Functions in the conductivity equation (2.20)

$E_{\Lambda,p}$ $E_{\Lambda,v}$ . . . . Constant pressure and constant volume activation energies of molar conductivity

$E_{\kappa,p}$ $E_{\kappa,v}$ . . . . Constant pressure and constant volume activation energies of conductivity

E_D . . . . Activation energy for diffusion

F_i . . . . Total force on ion i

F_j . . . . Total force on ion j

F_{12}^H . . . . Hard or repulsive force between particles 1 and 2

F_{12}^S . . . . Soft or attractive force between particles 1 and 2

$F(t)$ . . . . Instantaneous force on a particle at time t

$F(z)$ . . . . Function of z, equation (2.36)

G . . . . Gibbs function

$\Delta G_i^{\ddagger}$ . . . . Activation energy for ion i

$\Delta^{\ddagger}G_i^{\infty}$ .... Partial molar activation energy for conductance of
                     ion i at infinite dilution

ΔG_h. .... Gibbs energy of hole formation

H .... Solvation number of an ion

H .... Constant in equation (2.39) and has the value of
                     $3/8$ or $3/4$

H_0. .... Solvation number of an ion at infinite dilution

ΔH_h .... Enthalpy of hole formation

$\Delta H_i^{\ddagger}$. .... Activation enthalpy for conductance of ion i

$\Delta^{\ddagger}H_i^{\infty}$ .... Partial molar activation energy for conductance of
                     ion i at infinite dilution

J_1, J_2. .... Functions in the conductivity equation (2.20)

$J(t)$. .... Autocorrelation function of the total electric
                     current, equation (4.39)

K_R .... Association constant for solvent separated ion-pair
                     formation, equation (2.30)

K_S. .... Association constant for contact ion-pair formation

K_Λ .... Association constant obtained from the conductivity
                     equation (2.26)

L. .... Constant, equation (3.13)

L_{ij} .... $(i = \alpha, \beta, \gamma; j = \alpha, \beta, \gamma)$ The phenomenological
                     coefficient relating the flux of molecules i to the force
                     on molecules j

M .... Relative molecular mass of solvent, Chapter 2

N .... Number of subsystems, Chapter 3

N .... Number of molecules in a system, Chapter 4

N_A. .... Avogadro's number

N_0. .... Number of uncoordinated ions, equation (4.43)

N_x. .... Fraction of broken bonds, equation (3.18)

Q .... Partition function, Chapter 3

R. .... Radius of ion and cosphere, Chapter 2

S. .... Entropy

S. .... Constant in conductance equation (2.1)

S_c .... Configurational entropy

S_1, S_p .... Constants at 1 atm and at p atm

$\Delta S_i^{\ddagger}$. .... Activation entropy for conductance of ion i

$\Delta^{\ddagger}S_i^{\infty}$ .... Partial molar activation entropy for conductance of
                     ion i at infinite dilution

T. .... Temperature

T_g .... Experimental glass transition temperature; it is
                     usually about 20 K greater than T_0, and can be
                     obtained from heat capacity measurements

T_m. .... Melting temperature

T_{max} .... Temperature of conductivity maximum

T_0 .... Ideal glass transition temperature, equation (3.10)

U .... Internal energy

$U(r_{\alpha\beta})$ .... Pair potential for molecules α and β

$\Delta U_i^{\ddagger}$ .... Activation internal energy for conductance of ion i

$\Delta^{\ddagger}U_i^{\infty}$ .... Partial molar activation internal energy for conductance of ion i at infinite dilution

V .... Volume

V_f .... Free volume

V_h .... Hole volume

V_L .... Limiting volume

V_m .... Molar volume

V_S .... Volume of solid

V_0 .... Volume at T_0

V^* .... Critical free volume

ΔV_{Λ} .... Activation volume for molar conductivity

ΔV_{κ} .... Activation volume for conductivity

$\Delta V_i^{\ddagger}$ .... Activation volume for conductance of i

$\Delta^{\ddagger}V_i^{\infty}$ .... Partial molar activation volume for conductance of ion i

$\Delta^{\ddagger}V_j$ .... Activation volume for jumping of ion i into volume V_h

$W(T)$ .... Transition probability

X_1 .... Mean force acting on molecule 1, arising from a concentration gradient or an electric field

$X(t)$ .... Value of the fluctuating force X on a Brownian particle at time t

$Z_{\alpha}(t)$ .... Normalized velocity autocorrelation function for α ions

a .... Distance of closest approach of a cation and an anion and is obtained from the conductivity equation, e.g., equation (2.20)

a' .... Constant in equation (4.15)

c .... Concentration

$c_i, c_{\gamma}, c_{\alpha}, c_{\beta}$ ...... Concentration of ion i, γ, α, β, respectively

c .... Maximum close-packed coordination number, equation (4.46)

d .... Molecular jump distance (a molecular diameter)

d_i .... Half the jump distance of molecule i, see Fig. 4-4

e .... Charge of a proton

e_i .... $e_i = z_i e$, where $i = i, j, \ldots$

$f_{\pm}$ .... Mean ionic activity coefficient

f_{ij} .... Probability distribution function; the probability of finding a j ion at a distance $\mathbf{r}_{21}$ from an i ion at a distance $\mathbf{r}_2$ from an arbitrary origin

f_g .... Partition function of gaslike molecules

f^N .... Partition function of a system of N molecules

$f^{(N)}$ .... Phase space probability distribution function of N particles having position vectors $r_1, r_2 \ldots r_N$, and momenta $p_1, p_2 \ldots p_N$

$\bar{f}^{(n)}$ .... Time coarse-grained distribution function of $f^{(n)}$; the phase-space distribution function of n particles.

f_s .... Partition function of solidlike molecules

g .... Constant in equation (3.7)

$g(\tau_\kappa)$ .... Normalized density function of relaxation times

g_z .... Probability of occurrence of an ion with a coordination number z

$g^2(r_1, r_2)$ .... Pair correlation function; the probability of finding a molecule in the volume element dV_1 at r_1 if another molecule is in volume dV_2 at r_2 from the origin, normalized to 1 at large values of r.

$g^2(r_{\alpha\beta})$ .... Pair correlation function for molecules α and β separated by $r_{\alpha\beta}$

h .... Planck's constant

k .... Boltzmann's constant

k .... In equation (4.15), the transmission coefficient, from transition state theory

l_1, l_2, l_3 .... Dimensions of molecules; see Fig. 4-4

m .... Molality/mol kg^{-1}

n .... Number of molecules or ions in a subsystem

n_i .... Average density of i ions

n_{ij} .... Density of j ions at a distance r_{21} from an i ion at r_2 from the origin

p .... Pressure

p_i .... Momentum of particle i

q .... Constant in equation (3.7)

q_z^t .... Translational partition function of an ion with z neighbors

r .... Radius of an ion specified by the subscript, i.e., r_1, r_2, etc.

$r_{\alpha\beta}$ .... Distance between ions α and β

r_1 .... Position vector of ion i with respect to an arbitrary coordinate

u_i .... Electrical mobility of ion i

$u_{\alpha 1}$ .... Electrical mobility of ion $\alpha 1$

v .... Velocity of an ion

v_e .... Countercurrent velocity on an ion due to the electrophoretic effect

v_R . . . . Solvent flow about an ion due to asymmetry in the electric field around the ion

$\bar{v}_i$ . . . . Mean velocity of i ions

$\bar{v}_{\alpha 1}$ . . . . Mean velocity of ion $\alpha 1$

z . . . . In Chapter 3 the number of molecules in a subsystem

z^* . . . . In Chapter 3 the smallest number of molecules in a subsystem that permits a rearrangement

z . . . . Coordination number of an ion, equation (4.44)

z . . . . In Chapter 2 and equation (2.36), a function
$z = (S/\Lambda^{\infty 3/2})(c\Lambda)^{1/2}$

Greek letters

α . . . . In equation (2.46), the expansivity in equation (2.30), the fraction of paired ions present as contact ion pairs

α_0 . . . . In equations (2.1) and (2.12) the coefficient for the limiting relaxation term

β . . . . In equation (2.49) the isothermal compressibility

β_0 . . . . In equations (2.1) and (2.19) the coefficient for the limiting electrophoretic effect

γ . . . . In equation (2.30) the fraction of unpaired ions

γ . . . . In equation (3.6) the overlap of free volume factor

Δ . . . . Nernst–Einstein deviation parameter, equation (4.33)

ε . . . . Relative permittivity (dielectric constant)

ε_0 . . . . Relative permittivity at zero frequency

ε_∞ . . . . Relative permittivity at infinite frequency

ζ_0 . . . . Friction coefficient for a Brownian particle

ζ . . . . Friction coefficient for a microscopic particle

ζ^S . . . . Friction coefficient for the soft forces

ζ^H . . . . Friction coefficient for the hard forces

ζ_{ij} . . . . $(i = \alpha, \beta; j = \alpha, \beta)$ friction coefficient between ions i and j

η . . . . Shear viscosity of a liquid

η_s . . . . Shear viscosity for solidlike particles

η_g . . . . Shear viscosity for gaslike particles

η_0 . . . . Viscosity of pure solvent

θ . . . . Angle in Fig. 2-2 between the direction of the applied electric field and $\mathbf{r}_{21}$

κ . . . . Conductivity

κ . . . . In equation (2.13) the reciprocal of the κ^{-1} the radius of the "ionic atmosphere," the distance from the ion where the charge due to the ionic atmosphere reaches a maximum

κ_D . . . . Thermal diffusivity

κ^1 . . . . In equation (4.14) the frequency of jumping into a neighboring empty position

Λ . . . . Molar conductivity, except where stated otherwise, e.g., the equivalent conductivity

Λ_1 . . . . Molar conductivity at 1 atm pressure

Λ_p . . . . Molar conductivity at pressure p

Λ^∞ . . . . Molar conductivity at infinite dilution

$\Lambda(c)$ . . . . Molar conductivity at concentration c

λ_i . . . . Molar conductivity of ion i

λ_i^∞ . . . . Molar conductivity of ion i at infinite dilution

λ_E^∞ . . . . Excess molar conductivity at infinite dilution

ρ . . . . Density of a fluid

ρ . . . . In equation (4.28) the number density or number of particles per unit volume

σ . . . . In equation (4.11) the temperature-dependent surface tension

τ . . . . In equation (2.40) the dielectric relaxation time

τ . . . . In equation (2.13) a function defined by $\tau = e^2\kappa/2\varepsilon kT$

τ_κ . . . . Conductivity relaxation time

Introduction and Definitions

1.1. Introduction

Although the phenomenon of ionic conductance in simple liquids has been well characterised, the interpretation of the phenomenon is not well advanced. This is principally the result of our minimal knowledge of the molecular dynamics of ions in liquids. However, during the past 10 to 15 years, the results from relaxation spectroscopy and computer simulation of liquid particle dynamics have provided a new insight into the factors determining the mechanism of ionic transport. These factors are discussed in the chapters that follow.

Chapter 2 discusses the variation of conductance in dilute electrolyte solutions with concentration, temperature, and pressure. The classical problem of the concentration dependence continues to receive a lot of attention, and important developments have been made in recent years. High-speed computers have assisted the development because modern conductance equations are very complex, and it would be impractical to fit the data to them without computer assistance. Significant improvements have been made in our understanding of the factors that determine the limiting ionic conductivity in aqueous solutions. Transport number experiments, coupled with spectroscopic and thermodynamic measurements, have helped to identify some of the nonhydrodynamic contributions to ionic mobility. In aqueous solutions these contributions are largely from the water structure itself, and these seem to persist to quite high temperatures, pressures, and concentrations.

Molten salts and concentrated solutions are discussed in Chapters 3, 4, and 5. Aqueous solutions display a definite transition from dilute solution behavior to molten saltlike behavior, and high-pressure, high-temperature studies highlight this transition. Low-temperature molten salt transport characteristics seem best described by the configurational entropy theory, and a number of experiments have been carried out to establish the relationship of the glass transition temperature to transport phenomena. At higher temperatures well away from T_0, ionic melts can be simulated by fast computers. The results from these experiments have been used to test statistical mechanical theories of transport, as well as to provide a fresh insight into the factors that influence ionic mobility. Very high-pressure measurements of conductance in ionic melts are providing interesting results, the interpretation of which will challenge the theoreticians for some time to come.

At very high temperatures and low densities, molten salts display conductance maxima. This phenomenon is being characterised by a series of systematic researches, and our understanding of it is good, if only in qualitative terms.

The final chapter illustrates how the concepts of the previous five chapters can be applied to some geological and industrial electrolytes. As expected, these fluids are generally very complex, well beyond the concentration limit of classical dilute solution electrolyte theory, and are comprised of complex mixtures of solutes. Given the limitation of present theories as developed for simple systems, it is hardly surprising that conductance phenomena in complex systems can at best be described in qualitative terms.

This book seeks to provide a base from which an overall picture of the interpretation of conductance phenomena in liquids can be obtained. Where necessary, the reader is referred to other work for background material or, in many cases, for reading which is beyond the scope of this book.

1.2. Electrical Conductivity

The phenomenon of electric conduction arises from the movement of ions or electrons through the conducting system. Under the influence of an electric field E, the movement of ions gives rise to an *electric current I* defined as the rate of flow of charge,[1]

$$I = dq/dt \qquad (1.1)$$

where dq is the charge that passes through a cross-sectional area in time dt.† If A is the conductor's cross-sectional area, then the *current density j* is given by

$$j = I/A \qquad (1.2)$$

†One-dimensional motion is assumed.

and the *conductivity* (formerly *specific conductance*) κ by

$$\kappa = j/E \tag{1.3}$$

and

$$\rho = 1/\kappa \tag{1.4}$$

is called the *resistivity* of the conductor. When κ is independent of E, the conductor is said to obey Ohm's law. All the systems described in this book are thought to obey Ohm's law, at least at the low field strengths used for conductance measurements. The *resistance* R of the system is defined by

$$R = |\Delta\phi|/I$$

$|\Delta\phi|$ is the electric potential difference between the ends of a conductor of length l, and is called the *electric tension* (voltage). $|\Delta\phi|$ is related to the physical properties of the system by

$$|\Delta\phi| = Il/\kappa A$$

therefore

$$R = \rho l/A \tag{1.5}$$

The *conductance* of the system G is defined as

$$G = 1/R \tag{1.6}$$

R and G depend on the dimensions of the system and on its composition, whereas κ and ρ depend only on the composition.

If the conductance cell consists of two electrodes of area A and distance l apart, then the conductivity of the electrolyte is related to the measured resistance of the cell by

$$\kappa = l/AR \tag{1.7}$$

The ratio l/A is the *cell constant*; it has the dimensions of L^{-1} and the units of m^{-1}. However, it has been common practice (and apparently still is) to give the cell constant the units of cm^{-1}. Conductivity has the units of $S\ m^{-1}$ or $S\ cm^{-1}$, where S is the siemen, the S. I. unit for the reciprocal ohm, $1\ S = 1\ \Omega^{-1}$.

Under the influence of an applied electric field, the random motion of an ion will be sufficiently perturbed to produce a small component of acceleration in the direction of the field. A short time after the initial application of the electric field ($\sim 10^{-13}$ s), the electric and drag forces on the ion will balance, and the ion reaches its terminal velocity v_i. The drag forces experienced by the ion depend on its size and charge and the medium through which it moves, e.g., water, liquid NaCl, or molten SiO_2. An ion i of charge $z_i e$ under the influence of an electric field E will give rise to an electric current density j_i. From equations (1.1) and (1.2)

$$j_i = |z_i| e v_i N_i / Al$$

where N_i is the number of i ions in volume Al. If c_i is the concentration of i ions, and if all velocities are considered positive then

$$j_i = |z_i| F v_i c_i \tag{1.8}$$

The total current density due to all ions is therefore

$$j = \sum_i |z_i| F v_i c_i \tag{1.9}$$

From equation (1.3) the conductivity is

$$\kappa = \sum_i |z_i| F c_i \left(\frac{v_i}{E} \right) \tag{1.10}$$

Now at low field strengths κ is independent of E; therefore, the ratio (v_i/E) must be constant. This ratio is the velocity per unit field strength and is known as the electrical mobility u_i of ion i,

$$u_i = v_i / E \tag{1.11}$$

From equations (1.10) and (1.11)

$$\kappa = \sum_i |z_i| F c_i u_i \tag{1.12}$$

When only two types of ions, $+$ and $-$, are present, equation (1.12) and the condition of electroneutrality $\sum_i c_i z_i = 0$ give

$$\kappa = z_+ F c_+ (u_+ + u_-) \tag{1.13}$$

The *transport* number of an ion i is defined as the fraction of the current carried by that ion

$$t_i = j_i / j \tag{1.14}$$

Equations (1.3), (1.8), (1.11), and (1.14) give

$$t_i = |z_i| F c_i u_i / \kappa \qquad (1.15)$$

Thus the ionic mobility can be calculated once the transport number and the conductivity have been experimentally determined. The ionic mobility is an invaluable aid to the interpretation of electrolyte conductivities because it enables the physical chemist to evaluate the contribution due to each ion and how that contribution varies with concentration, solvent, pressure, and temperature.

The *molar conductivity* of an electrolyte is the conductivity per mole of solute, Λ_m, where

$$\Lambda_m = \kappa / c \qquad (1.16)$$

If the units of κ are $S\ m^{-1}$ and the units of concentration are mol m^{-3}, then Λ_m has the units of S m^2 mol^{-1}. However, since κ is often expressed in S cm^{-1}, then the concentration must be expressed in mol cm^{-3} to give Λ_m in units of S cm^2 mol^{-1}.

From equations (1.13) and (1.16) it can be seen that Λ_m is directly proportional to the sum of the ionic mobilities in a given system. The variation of the molar conductivity with concentration, temperature, and pressure gives a direct insight into the effect of these variables on ionic mobility and, as such, is a more useful property than the conductivity κ.

The molar conductivity of an ion is defined as

$$\lambda_{m,i} = \kappa_i / c_i \qquad (1.17)$$

where c_i is the concentration of the ion i, and κ_i is the contribution of the i ion to the conductivity. $\lambda_{m,i}$ is related to Λ_m by

$$\Lambda_m = \frac{1}{c} \sum_i c_i \lambda_{m,i} \qquad (1.18)$$

and to the ionic mobility of ion i by

$$\lambda_{m,i} = |z_i| F u_i \qquad (1.19)$$

For various reasons, some authors[2] express conductivity in terms of molality (mol kg^{-1}), thus the *molal conductivity* is defined as

$$\Lambda_m = \kappa / m$$

where Λ_m has units of S kg cm^{-1} mol^{-1}.

The *equivalent conductivity* Λ_{eq} is the conductivity per equivalent of solute. Thus Λ_{eq} of the solute $M_{\nu_+}^{z_+} X_{\nu_-}^{z_-}$ is defined as

$$\Lambda_{eq} = \frac{\kappa}{\nu_+ |z_+| c} \qquad (1.20)$$

However, the continued use of equivalent conductivity, along with the concept of equivalents and normalities is not recommended.[3] It is defined in this book solely because much of the existing data is quoted in these terms.

1.3. Diffusion and Viscosity

1.3.1. The Diffusion Coefficient

If J_i is the flux of i ions arising from a gradient of chemical potential $\partial \mu_i / \partial x$, then J_i may be written as

$$J_i = L\left(-\frac{\partial \mu_i}{\partial x} \right) \qquad (1.21)$$

where L is a phenomenological coefficient. J_i is the number of ions crossing a unit area normal to the direction of flow per unit time; therefore,

$$J_i = c_i v_i \qquad (1.22)$$

Now the force $-(\partial \mu_i / \partial x)$ on the ion must be balanced by an equal and opposite force, otherwise the ion would continue to accelerate. If the opposing force is regarded as a drag force F_d and is proportional to the velocity of the ion, then

$$-\frac{\partial \mu_i}{\partial x} = -F_d = \zeta v_i \qquad (1.23)$$

where ζ_i is a drag coefficient. When the concentration of ions is small, equations (1.22) and (1.23) give

$$J_i = \frac{-c_i}{\zeta} \frac{\partial \mu_i}{\partial x} = \frac{-kT}{\zeta} \frac{\partial c_i}{\partial x}$$

or

$$J_i = -D_i \frac{\partial c_i}{\partial x} \qquad (1.24)$$

where the diffusion coefficient

$$D = \frac{kT}{\zeta} \tag{1.25}$$

Equation (1.24) is known as Fick's first law of diffusion; D_i is a constant for a given system at constant temperature and pressure.

1.3.2. Viscosity

The viscosity η of a Newtonian fluid is defined by the ratio

$$\eta = -\sigma_{xy}/v_{xy} \tag{1.26}$$

where σ_{xy} is the shear stress in the x direction exerted by the fluid on a surface of unit area which is normal to the y direction. v_{xy} is the velocity gradient whose directional properties are similarly indicated by the suffixes, the x component of velocity varying in the y direction. For a Newtonian fluid η is a constant, but for a non-Newtonian fluid it is a function of the applied shear stress. Non-Newtonian fluids are generally polymer solutions, or fluids that contain a solid phase; it is believed that the liquids referred to in this book are Newtonian. However, since conductance measurements are far easier to make than viscosity measurements the precision and the pressure and temperature range covered by conductance studies far exceeds that of viscosity measurements. It is possible that some of the liquids referred to in this book become non-Newtonian at high pressures.

1.4. The Stokes–Einstein and Nernst–Einstein Relations

1.4.1. The Stokes–Einstein Relation

According to Stokes' law the hydrodynamic force needed to move a sphere of radius r, with a velocity v, through a continuum of viscosity η is

$$F = 6\pi\eta r v \tag{1.27}$$

However, the constant 6 applies only in the case where the fluid molecules adhere perfectly to the surface of the sphere. On the other hand, when the sphere can move through the fluid with no adhesion the constant has the value of 4.[4] Equation (1.27) provides in principle a way of evaluating the friction coefficient in equation (1.23). If it is assumed that Stokes' law is valid for microscopic particles of similar size to solvent molecules, then the hydrodynamic drag force on an ion moving under the influence of a chemical potential gradient is from equations (1.23) and (1.27):

$$- F_d = 6\pi\eta r_i v_i = \zeta v_i \tag{1.28}$$

When combined with (1.24) this gives

$$D_i = \frac{kT}{6\pi\eta r_i} \tag{1.29}$$

the Stokes–Einstein equation. The Stokes–Einstein equation can only be regarded as an approximate relation that becomes more reliable as the radius of the diffusing ion becomes greater than that of the solvent molecules. In many applications of this relationship the constant 6π is regarded as an adjustable parameter that varies between 4π and 6π.

1.4.2. The Nernst–Einstein Relation

This equation provides an approximate relationship between the ionic molar conductivity and the ionic diffusion coefficient. The force F_E, acting on an ion due to the electric field is

$$F_E = z_i e E \tag{1.30}$$

If this force is opposed by an equal and opposite drag force given by

$$F_E = - F_d = \zeta v_i$$

then from equations (1.30), (1.25), and (1.11)

$$\frac{u_i}{e} = \frac{z_i D_i}{kT} \tag{1.31}$$

and from equation (1.19)

$$\frac{\lambda_{m,i}}{F^2} = \frac{|z_i|^2 D_i}{RT} \tag{1.32}$$

In terms of the molar conductivity of a solute composed of $+$ and $-$ ions

$$\frac{\Lambda_m}{F^2} = \frac{|z_+|^2 D_+ + |z_-|^2 D_-}{RT} \tag{1.33}$$

For molten salts the ratio of the calculated molar conductivity to the measured value is always greater than unity, and reasons for this are discussed in Chapter 4. An equivalent equation called the Nernst–Hartley relation has been developed for electrolyte solutions; it is valid at infinite dilution only.

1.4.3. The Walden Product

If the electric force on an ion is equated to the hydrodynamic drag force as given by Stokes' law, then it can be shown that

$$\lambda_{m,i}\eta = \frac{\text{constant}}{r_i} \tag{1.34}$$

The product $\lambda_{m,i}\eta$ is called Walden's product and for a given solvent viscosity should be inversely proportional to the ionic radius. Deviations from this rule are discussed in some detail in Chapter 2.

Ionic Conductivity in Dilute Electrolyte Solutions

2.1. Concentration Dependence of Conductivity of Dilute Electrolyte Solutions—Introduction

Dilute electrolyte solutions will be regarded as those which fall into the concentration range where the classical Debye–Hückel model of electrolyte solutions is valid. This model regards ions in an electrolyte as being hard spheres separated by a dielectric continuum whose relative permittivity, ε, is that of the pure solvent. In water this model is valid up to $0.001 \text{ mol dm}^{-3}$ for $1:1$ electrolytes.

The conductivity curves for KCl in water–dioxane mixtures, Fig. 2-1, provide a typical example of the dependence of Λ on concentration and relative permittivity.[1] Curve F is for KCl in water and lies above the limiting tangent as do all conductivity curves for strong electrolytes in solvents with a high relative permittivity. The expression for the limiting tangent,

$$\Lambda = \Lambda^\infty - Sc^{1/2} = \Lambda^\infty - (\alpha_0\Lambda^\infty - \beta_0)c^{1/2} \qquad (2.1)$$

where Λ^∞ is the conductivity in the absence of all interionic effects, was discovered empirically by Kohlrausch[2] and proved theoretically by Onsager[3] in 1926. It predicts the slope of the $\Lambda-c^{1/2}$ curves at infinite dilution. For values of $\varepsilon < 41.46$ the conductivity curves lie below the limiting tangent due to the effects of ion association; as ε decreases the curves become progressively steeper and assume the sigmoid shape typical of highly associated electrolytes.

Ever since Kohlrausch and co-workers discovered their empirical equation, it has been the task of theoreticians and experimentalists working in the field to establish an equation that accounts for the variation of Λ with c, ε, electrolyte, and solvent. Most work has been concerned with $1:1$ electrolytes and is described in detail in texts and review articles.[4] Of particular note are: *Electrolytic Conductance* by Raymond M. Fuoss and

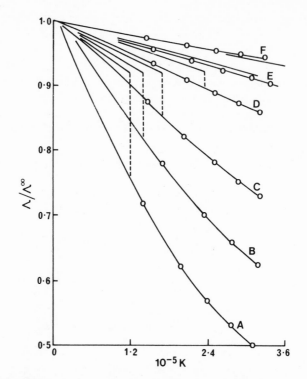

FIG. 2-1. Normalized conductance curves for potassium chloride in water–dioxane mixtures for the range $78.54 > \varepsilon > 12.74$. (A, 12.74; B, 15.37; C, 19.32; D, 30.26; E, 41.46; F, 78.54). $\kappa = \left(\dfrac{4\pi N_A e^2}{1000 \varepsilon kT} \sum_i c_i \gamma_i z_i^2 \right)^{1/2}$. Reprinted with permission from R. M. Fuoss, *Rev. Pure and Appl. Chem.* **18**, 125–136 (Fig. 1) (1969).

Fillipo Accascina[5]; *Electrolyte Solutions* by R. A. Robinson and R. H. Stokes[106]; and *Theorie der Elekrolyte* by H. Flakenhagen, W. Ebeling, and H. G. Hertz.[107] These texts cover this subject in considerable detail and the interested reader is referred to them for the detailed derivation of conductance equations and analysis of results up to 1970. There have recently been a number of improvements, however, to conductance theories of dilute solutions, and they will be discussed in outline below, along with a brief discussion of their derivation.

2.1.1. Derivation of the Conductance Equations

The primitive model upon which conductance theories are based is that of a hard sphere ion of radius a and charge $z_i e_i = e_i$, drifting in a continuum of relative permittivity ε and viscosity η, under the influence of an applied electric field E.[5] The reduction in mobility as concentration is increased from infinite dilution is due to the interionic effects of electrostatic interaction through the coulomb potential. Dominant are the relaxation and electrophoretic effects and their cross terms.

An ion of charge e_i is surrounded in solution by an atmosphere of ions whose net charge is $e_j = -e_i$; according to the Debye–Hückel theory this charge is maximum at a distance of $1/\kappa$ from the central ion and is referred to as the "ionic atomsphere" of the ion. When an external electric field is applied to the electrolyte solutions, the central ion will be attracted toward an electrode. Consequently, the previously spherically symmetric field around the central ion becomes asymmetric because the ionic atmosphere cannot "relax" fast enough so as to adjust to the new position of the central ion. This creates a small electrostatic braking force on the central ion, and its mobility is reduced below the value at infinite dilution. If the external field acting on the ion produces a force of $e_i E$, where E is the field intensity, then we represent the relaxation force by $-\Delta E_R$. Thus the mobility of the ion is reduced by a factor $1 - \Delta E_R/E$.

When an ion moves in an electrolyte solution it tends to drag the local solvent molecules with it. Since anions and cations travel in opposite directions under the influence of an external electric field, then every ion will be moving against a stream of solvent molecules. This is the electrophoretic effect, and it gives rise to a countercurrent velocity v_e. There are, of course, cross terms that arise from the interaction of the above effects. The hydrodynamic contribution to the relaxation term arises from the velocity field produced in the solvent at the site of an i ion, by the movement of a nearby j ion. This flow contributes to the asymmetry of the potential surrounding an ion, and hence contributes a term ΔE_v to the relaxation force. Conversely, an asymmetry in the field around an ion caused by the external field will cause an additional solvent flow about the ion and hence a term v_R.

A further term is often included, and this arises from the asymmetry in the distribution of j ions about an i ion caused by countercurrent ion migration. Because of this asymmetry and mutual ion attraction, an i ion is more likely to be struck by a j ion from behind than by one in front. This effect is equivalent to a small force that acts in the same direction as the external field: it has been referred to as the "osmotic field", ΔE_0.

Precise conductivity equations also take into account the effect of ionic volume on the viscosity of the solvent, by dividing the conductivity equation by $1 + Bc$. B is the Jones–Dole viscosity coefficient, or a constant derived from Einstein's equation for the viscosity η, of a solution whose solvent viscosity is η_0 and whose solute volume fraction is ϕ,

$$\eta = \eta_0 \left(1 + \frac{5}{2} \sum_{i=1}^{j} \phi_i \right)$$

where

$$\phi_i = \left(\frac{4 \pi r_i^3}{3} \right) \left(\frac{Nc_i}{1000} \right) = \frac{2 B_i c_i}{5}$$

r_i is the hydrodynamic radius of i, c_i is the concentration.

The total effect of all these perturbations on the limiting ionic mobility can be summarized into one equation for the molar conductivity of a binary electrolyte:

$$\Lambda = \sum_{i=1}^{2} |z_i| \left[\lambda^\infty \left(1 - \frac{\Delta E_R}{E} - \frac{\Delta E_v}{E} + \frac{\Delta E_0}{E} \right) - \frac{F(v_R + v_e)}{E} \right] \Big/ (1 + B_i c_i)$$

(2.2)

The forces on the ions resulting from the applied field plus interionic effects, and their conjugate fluxes, may also be described in terms of the principles of nonequilibrium thermodynamics.[6] If J_i is the flux of the i ion in ions $cm^{-2} s^{-1}$, resulting from the forces $e_i E, e_j E$, then

$$J_i = \Omega_{ii} e_i + \Omega_{ij} e_j$$

$\Omega_{ii} = n_i \omega_i$ is the phenomenological coefficient, where n_i is the number density of i ions and ω_i is the mobility of i etc. In accordance with Onsager's reciprocity relation, Chen has shown

$$\Omega_{ij} = \Omega_{ji}$$

The reciprocal terms Ω_{ij}, Ω_{ji} are a measure of the modification of the ion mobility of an i ion by interaction with an oppositely charged j ion; Ω_{ji} contains terms from both the electrophoretic and relaxation effects and their cross terms.

The derivation of explicit expressions for the relaxation and electrophoretic terms in (2.2) is very complex; a brief discussion of the method will be given here, with particular reference to the physical approximations that are made.

2.1.2. The Relaxation Effect

At equilibrium, i.e., when $E = 0$, the potential about the position of an i ion is spherically symmetric. However, when an electric field is applied a small asymmetry develops in the potential field ψ'; this is related to the relaxation field by

$$\Delta E_R = -\left(\frac{\partial \psi'}{\partial x}\right)_a \tag{2.3}$$

where the applied field is in the x direction. ψ' is calculated by integrating the Onsager continuity equation,[3] subject to suitable boundary conditions. The continuity equation can be written as

$$\nabla_1 \cdot (f_{ij}\mathbf{v}_{ij}) + \nabla_2 \cdot (f_{ji}\mathbf{v}_{ji}) = 0 \tag{2.4}$$

where f_{ij}, the distribution function, is defined by

$$f_{ij}(\mathbf{r}_2, \mathbf{r}_{12}) = n_i n_{ij}(\mathbf{r}_2, \mathbf{r}_{12}) \tag{2.5}$$

n_i is the average density of i ions, n_{ij} is the density of j ions at a distance $\mathbf{r}_{21}$ from an i ion at $\mathbf{r}_2$ from an arbitrary origin, Fig. 2-2. In the presence of an applied field

$$f_{ij} = f_{ij}^0 + f_{ij}^1 \tag{2.6}$$

where $(f_{ij} = f_{ij}^0)_{E=0}, f_{ij}^1$ represents the asymmetric part of the distribution

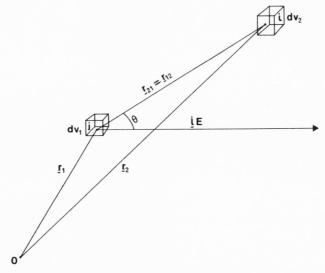

FIG. 2-2. Definition of distribution function.[4a]

function; similarly

$$\psi_j(\mathbf{r}_1\mathbf{r}_{21}) = \psi_j^0(r) + \psi_j^1(\mathbf{r}_1, \mathbf{r}_{21}) \tag{2.7}$$

where $\psi_j(\mathbf{r}_1, \mathbf{r}_{21})$ is the potential at an i ion $\mathbf{r}_{21}$ from a j ion at $\mathbf{r}_1$. ψ_j can be related to (2.6) by the Poisson equation

$$\nabla^2 \psi_j = -4\pi/\varepsilon n_j \sum_i e_i f_{ij} \tag{2.8}$$

In (2.4) $\mathbf{v}_{ji}$ is the velocity of an i ion at a distance $\mathbf{r}_{21}$ from a j ion; it is the sum of $\mathbf{v}_i$, the velocity with respect to a stationary observer of the solvent at the location of the i ion, and the velocity produced by the forces acting on the ion, viz.,

$$\mathbf{v}_{ji} = \mathbf{v}_i + \zeta_i(\mathbf{K}_{ji} - kT\nabla_2 \ln f_{ji}) \tag{2.9}$$

ζ_i is the friction coefficient of the i ion, $kT\nabla_2 \ln f_{ji}$ is the diffusion force which opposes the external field and arises from f_{ji}^1. K_{ji} is the total electrostatic force on the i ion and is comprised of three terms,

$$\mathbf{K}_{ji} = Ee_i\hat{\mathbf{i}} - e_i\nabla_2\psi_i^1(a) - e_i\nabla_2\psi_j(\mathbf{r}_1, \mathbf{r}_{21}) \tag{2.10}$$

$Ee_i\hat{\mathbf{i}}$ is the force in the direction of the applied field, the second term is the force at the surface of the ion arising from the asymmetry in the ion atomsphere, the last term arises from the presence of a neighbor e_j and its ionic atomsphere. Equations (2.4), (2.6)–(2.10) give, subject to

    a.   the choice of f_{ji}^0,
    b.   suitable boundary conditions to (2.4),
    c.   the mathematical approximations used in the integration of (2.4),

a solution to equation (2.3).

The simplest form of f_{ji}^0 is a Boltzmann distribution,

$$f_{ji}^0 = n_j n_i \exp(-e_i\psi_j^0/kT) \tag{2.11}$$

which for symmetrical electrolytes is expanded to

$$f_{ji} = n_j n_i\left[1 - e_i\psi_j^0/kT + \tfrac{1}{2}(e_i\psi_j^0/kT)^2\right]$$

Other more complex distribution functions are discussed in the texts referred to in the Introduction. Four boundary conditions are needed to evaluate the four integration constants obtained from (2.4). The conditions are chosen to suit the physical properties of the model and differ slightly from author to author.

The first condition is based on the fact that the influence of the central j ion, regarded as being at the origin where $r = 0$, is negligible at $r = \infty$. Second and third conditions are that the potential and field gradient should be continuous at $r = a$, the surface of the ion. These three conditions are common to several groups of researchers,[7-10] but the fourth conditions has been chosen in two mutually incompatible ways. The schools of Fuoss and of Falkenhagen have chosen the condition that the relative velocity of two ions along the line of centers must be zero at $r = a$; this implies that f_{ij}^1, f_{ji}^1 are nonzero at $r = a$, while Pitts and Carman have preferred the condition that f_{ij}^1, f_{ji}^1 are zero at $r = a$. However, according to Carman, the two conditions give essentially the same result at low values of concentration, and he, therefore, prefers the latter because it eases the mathematical complexity in solving the continuity equation.

Recently Fuoss[11] has presented an improved model for ionic conductivity in dilute solutions. Because of the importance of this development it will be described in more detail later in this section, but a brief description of the boundary conditions will be given here because the following discussion refers frequently to this model. The model is one of "Gurney cospheres" of radius R in a continuum of relative permittivity ε and viscosity η. Equation (2.4) is solved according to the conditions of electroneutrality, continuity of the potential, and field gradient at $r = R$, that an unpaired ion contains no other ion in its cosphere, and that the asymmetry in the distribution function vanishes for ions at infinite separation. The approximation solution for the electrostatic contribution to the relaxation field can be expressed as the sum of terms

$$\frac{\Delta E_R}{E} = \sum_{i=0} \frac{\Delta E_i}{E}$$

where each ΔE_i represents increments to solution of the continuity equation to higher orders of c, e.g., ΔE_0 is to order $c^{1/2}$, ΔE_1 to order $c \log c$, ΔE_2 to c, c^2, $c^{3/2} \ln c$, etc. $\Delta E_0 / E$ is given by

$$\frac{\Delta E_0}{E} = - \frac{\tau}{3(1+y)(1+t)(1+yt)}$$

where

$$\tau = \frac{e^2 \kappa}{2\varepsilon kT} = \frac{\beta \kappa}{2}$$

and

$$t = \kappa R, \qquad \kappa = \left(\frac{4\pi N_A e^2}{1000\varepsilon kT} \sum_i c_i \gamma_i z_i^2 \right)^{1/2}, \qquad y = \left[\frac{z_i \omega_i + z_j \omega_j}{(z_i + z_j)(\omega_i + \omega_j)} \right]^{1/2}$$

γ_i is the fraction of unpaired ions. In the limit of low concentrations

$$\frac{\Delta E_{OL}}{E} = -\frac{\tau}{3(1+y)} = -\alpha_0 c^{1/2} \tag{2.12}$$

the classical Onsager result for the limiting relaxation term in equation (2.1).

The succeeding terms become very complex, but the sum of the relaxation terms that arise solely from electrostatic effects can be represented by

$$\frac{-\Delta E_R}{E} = \frac{\Delta E_0}{E} + \frac{\tau^2}{3}\ln t + \beta^2 \kappa^2 (FH2) + \beta^3 \kappa^3 (FH3) \tag{2.13}$$

where, for computation purposes, the $(FH2)$, $(FH3)$ can be represented as cubic polynomials in t.

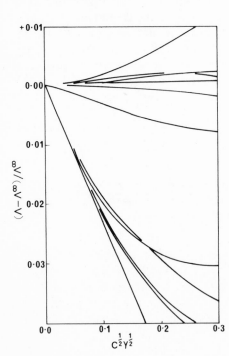

FIG. 2-3. Dependence of relaxation terms on $(c\gamma)^{1/2}$ for CsBr in water. Top curves $\Delta E_0/E$; curves terminating at $(c\gamma)^{1/2} = 0.3$ in sequence from top $\Delta E_{24}/E$, $\Delta E_7/E$, $\Delta E_8/E$, $\Delta E_3/E$, $\Delta E_1/E$, total $\Delta E/E$, $\Delta E_0/E$; lower curves $\Delta E_R/E$, $(\Delta E_0/E + \Delta E_1/E)$; straight line $-\alpha_0(c\gamma)^{1/2}$. Note that $(\Lambda - \Lambda^\infty)/\Lambda^\infty = \Delta E_i/E$. Reprinted with permission from R. M. Fuoss, *J. Phys. Chem.* **82**, 2427–2440 (Fig. 1) (1978). Copyright 1978 American Chemical Society.

The effect of the relaxation terms $\Delta E_i/E$ on the fractional change in the calculated value of Λ is shown in Fig. 2-3. The minimum in the total $\Delta E/E$ versus $(c\gamma)^{1/2}$ curve does not, however, produce a minimum in the Λ versus $(c\gamma)^{1/2}$ plots since the fractional change in Λ due to the electrophoretic effect is larger than the relaxation effect. Therefore, their sum, equation (2.2), decreases continuously with increasing concentration (Fig. 2-4). That $\Delta E_R/E$ should pass through a minimum is also predicted by Carman. In this work Carman[10] has produced a very accurate solution for the relaxation effect by computer assisted numerical analysis. His results show that for 1:1 electrolytes in water $-\Delta E_R/E_R$ decreases rapidly after passing through a small maximum at $\sim 0.19 \text{ mol dm}^{-3}$ and is always considerably smaller than the electrophoretic contribution.

2.1.3. The Electrophoretic Effect

The force on an ion due to the external electric field is transmitted to solvent molecules. Thus, from Stokes' equation, the force density $E\rho\hat{\mathbf{i}}$ on a spherical shell of radius r and thickness dr, imparts a velocity in the field direction $\hat{\mathbf{i}}$ of

$$d\mathbf{v}_e = (E\rho\hat{\mathbf{i}})4\pi r^2 \, dr/6\pi\eta r \qquad (2.14)$$

to an ion at the center of the shell. The charge density, ρ, is given by the Poisson equation

$$\nabla^2\psi^0 = \frac{4\pi}{\varepsilon}\rho(r) \qquad (2.15)$$

and the total electrophoretic velocity produced by all the ions in the ionic atmosphere of the center ion is obtained by integrating from a to ∞. For symmetrical electrolytes the result due to Fuoss and Onsager[9] is

$$\mathbf{v}_e = -\frac{Ee\kappa\hat{\mathbf{i}}}{6\pi\eta(1+\kappa a)} \qquad (2.16)$$

However, it has been shown recently that (2.16) is inadequate.[12, 13] The calculation should involve the total force K_{ji} acting on an ion and not just that due to the external field. The evaluation of the electrophoretic term becomes somewhat more complex but remains similar in principle. The contribution to the velocity of the solvent in the field direction $d\mathbf{v}_e$, at the location of a j ion at the origin, due to a force $d\mathbf{F}_j$ acting on a volume element dV also at the origin is

$$d\mathbf{v}_e = \left[\hat{\mathbf{i}} \cdot d\,\mathbf{F}_j + (\hat{\mathbf{r}} \cdot d\,\mathbf{F}_j) \cos\theta\right] / 8\pi\eta r \qquad (2.17)$$

$\hat{\mathbf{r}}$ is the unit vector in the radial direction r_{ij} and θ is the angle between $\hat{\mathbf{i}}$ and $\hat{\mathbf{r}}$. The total force on the j ion is the force on the volume element of solvent, which is the force on the ions in dV, i.e.,

$$d\,\mathbf{F}_j = dV \sum_i n_{ji} \mathbf{K}_{ji} \qquad (2.18)$$

$\mathbf{K}_{ji}$ is the force acting on an i ion located r_{ji} from j and is given by (2.10). When integrated over θ and from $r = R$ to $\propto$, the result obtained by Fuoss' model[11] for the total electrophetic effect is

$$v_e = \frac{\beta_0 c^{1/2} \gamma^{1/2}}{1 + t} + \text{higher terms in } t \qquad (2.19)$$

where $\beta_0 = E e_j \kappa / 6\pi\eta c^{1/2}$. The higher terms that contain the cross term v_R make a significant contribution to the total Λ at higher concentrations and reduce the effect of countercurrent solvent flow. These terms can also be expressed, for analytical purposes, in the form of polynomials in t.

The electrophoretic contribution to the relaxation effect $\Delta E_V / E$ requires the calculation of the radial components v_{ir}, v_{jr} or v_i, and v_j, and they are evaluated in a similar manner to that described above for the principal electrophoretic term. As before, this term can be expressed in terms of a polynomial in t.

2.1.4. Conductivity Equations for Nonassociated Electrolytes

The combined results from the approximate solution to the continuity equation (2.4) and the expressions for the electrophoretic effect and cross terms, have been expressed in the form[5, 14, 16]

$$\Lambda = \Lambda^\infty - S c^{1/2} + (E_1 \Lambda^\infty - E_2) c \ln c + J_1 c + J_2 c^{3/2} \qquad (2.20)$$

The first two terms on the right-hand side constitute the Onsager limiting law, where S is a function of κ, ε, T, η and arises from terms to the order of $c^{1/2}$ in the relaxation and electrophoretic effects. E_1 is a function of κ, ε, T, and it arises from terms to the order of $c \ln c$ in the expression for the relaxation effect. The second term in $c \ln c$, E_2, arises from the electrophoretic contribution to the relaxation field, $\Delta E_V / E$ and its conjugate term v_R. Earlier workers missed the latter effect, and the E_2 term in their papers must be multiplied by 2 to give the correct value for the total $c \ln c$ coefficient.[13] The expressions for J_1 and J_2 arise from higher terms in the solution of the continuity equation for the relaxation field and are complex functions of ε, κ, T, η and the ion size parameter a. These coefficients, as given in the equations due to Onsager and Fuoss, Pitts, and Falkenhagen and Kremp, are compared in the review by Barthel.[4a]

More recently, improved solutions to the continuity equation have proved to be so complex that it becomes impractical to express the equation in the form of equation (2.20). The conductance equation must be left in a form similar to (2.2), with appropriate values for the contributing terms calculated from the explicit expressions or from interpolating functions.

2.1.5. Conductivity Equations for Associated Electrolytes

The conductivity equations (2.2) and (2.20) can be modified to account for ion association by the *ad hoc* hypothesis that the concentration of ion pairs is given by the mass action equation

$$1 - \gamma = K_\Lambda c \gamma^2 f_\pm^2 \tag{2.21}$$

and that ion pairs make no contribution to the conductivity.[15] Given these assumptions equation (2.20) becomes, for a $1:1$ electrolyte

$$\Lambda(1 + Bc) = \gamma \left[\Lambda^\infty \left(1 - \frac{\Delta E}{E} \right) + \frac{F v_e}{E} \right] \tag{2.22}$$

or in expanded form

$$\Lambda = \gamma \left[\Lambda^\infty - S(\gamma c)^{1/2} + E'(\gamma c) \ln(\gamma c) + J_i(\gamma c) + J_2(c^{3/2} \gamma^{3/2}) \right] \tag{2.23}$$

The expressions chosen for the activity coefficient $f_\pm$ differ; Fuoss in earlier papers chooses the Debye–Hückel limiting law, whereas other authors use the full Debye–Hückel expression[15, 16]

$$\log f_\pm = \frac{-e_i e_j \kappa}{2\varepsilon k T(1 + \kappa a)} \tag{2.24}$$

If ion association is a consequence of coulomb forces only, then the corresponding decrease in conductivity with increasing concentration should be predictable from the theory. Fuoss *et al*[17] have shown that by retaining the complete Boltzmann factor in the distribution function, an equation of the functional form of (2.23) can be justified theoretically,[17] viz.,

$$\Lambda = \Lambda^\infty - Sc^{1/2}\gamma^{1/2} + E'(c\gamma)\ln(6E'c\gamma) + Lc\gamma - K_\Lambda c\gamma f_+^2 \Lambda \quad (2.25)$$

where K_Λ is given by equation (2.21) and is identified with the association constant for pairs of ions in contact.

2.1.6. Test of the Conductivity Equation

Equations (2.22) and (2.23) can be written symbolically as

$$\Lambda = \Lambda(c, \Lambda^\infty, K_\Lambda, a) \quad (2.26)$$

where Λ^∞, K_Λ, and a are unknown. In order to test the fit of conductivity data to a conductivity equation, it is therefore necessary to find appropriate values of Λ^∞, K_Λ, and a that minimize the standard deviation δ_Λ, between the observed and calculated conductivities. The data used in such determinations are generally precise to 0.01%, and equations of the form of (2.22) and (2.23) can be fitted with a standard deviation ranging from 0.005% of Λ^∞ to 0.2% over many salt–solvent systems.[15]

The magnitude of the parameters derived from a conductivity equation provides a test for the equation as well as being of intrinsic interest. The dependence of Λ^∞ on ion size, solvent structure, temperature, and pressure is fundamental to any discussion of conductance mechanisms and is discussed in detail in Section 2.3. However, Λ^∞ is not sensitive to the equation from which it was derived, and this is not surprising since all conductivity equations become (2.1) in the limit of extreme dilution. Table 2-1 lists some Λ^∞ values determined from different equations; note that close to the minimum value of δ_Λ, Λ^∞ seems to be independent of both K_Λ and a. K_Λ is, however, sensitive to the form of the conductance equation. Furthermore, a range of (K_Λ, a) pairs, giving almost equally good fit to a conductivity equation, can be found, e.g., AgNO$_3$, Table 2-1. In this Table (FO) refers to the 1957 Fuoss–Onsager equation[9] where $J_2 = 0$, (FHFP) an expansion of the Fuoss and Hsia (FH)[18] equation by Fernandez-Prini[19], (PFPP) and expansion of the Pitts[8] equation by Fernandez-Prini and Prue[14], and (FJ) a form of the Fuoss–Hsia equation modified by Justice[20] so that a is identified as q, where

$$q = \frac{\beta}{2} = \frac{z^2 e^2}{2\epsilon kT} \quad (2.27)$$

It is evident from the table that even for a given equation the value of K_Λ is dependent on the choice of a.

TABLE 2-1. Parameters Derived from Conductivity Equations

Equation	Electrolyte	Solvent	ε	Λ^∞/S cm² mol⁻¹	K_Λ/mol⁻¹ dm⁻³	a/Å	δ_Λ	Sum of ionic radii in Å	q/Å	Reference
FJ	Bu$_4$NBr	Acetonitrile	36.02	162.16	13.2	7.4	0.02		7.78	102
FO	Bu$_4$NBr	Acetonitrile	36.02	162.1	2	3.5				103
FJ	Bu$_4$NBPh$_4$	Acetonitrile	36.02	119.66	5.1	7.3	0.02		7.78	102
FO	Bu$_4$NBPh$_4$	Acetonitrile	36.02	119.49		5.3				103
PFPP	CsI	Water 22.1 wt%	60.18	97.39	0.00	3.49	0.023	3.87	4.66	23
FJ/FHFP	CsI	Dioxane	60.18	99.40	0.5	4.52	0.02		4.66	23
PFPP	CsI	Water 57.1 wt%	29.79	56.38	8.0	3.69	0.027		9.41	23
FJ/FHFP	CsI	Dioxane	29.79	56.51	20.8	8.03	0.003		9.41	23
			29.79	56.49	13.8	5.99	0.005			23
PFPP	CsI	Water 70.7 wt%	18.68	48.06	119	4.71	0.061		15.0	23
FJ/FHFP	CsI	Dioxane	18.68	48.49	187	11.16	0.004		15.0	23
			18.68	48.39	145	7.63	0.008			23
PFPP	CsI	Water 78.5 wt%	12.81	42.19	1054	5.73	0.033		21.87	23
FJ/FHFP	CsI	Dioxane	12.81	42.02	937	7.11	0.009		21.87	23
			12.81	42.69	1362	18.86	0.022			23
FO	KCl	Water	78.35	149.89	0.711	5.65	0.01	3.14	3.58	104
FH	KCl	Water	78.35	149.95	0.026	3.53	0.01		3.58	104
FH	KNO$_3$	Water	78.35	145.03	0.597	3.42	0.02	3.8	3.58	104
Carman	KNO$_3$	Water	78.35	145.05	1.4	4.22	0.0268			105
Carman	AgNO$_3$	Water	78.35	133.35	0.9	2.74	0.0386	3.8	3.58	105
				133.36	1.0	3.60	0.0382			105
				133.36	1.1	3.46	0.0378			105
				133.36	1.2	3.82	0.0377			105
				133.36	1.3	4.21	0.0378			105
				133.36	1.5	4.99	0.0382			105
Carman	LiCl	Water	78.35	114.97	0.1	3.15	0.0464	3.3[a]	3.58	105
				114.97	0.2	3.51	0.0464			105
Carman	NaF	Water	78.35	105.74	0.7	3.68	0.0333	3.7[b]	3.58	105
				105.74	0.8	4.03	0.0333			105

[a] Calculation assumes r_{Li^+} is hydrated and equal to 1.5 Å.
[b] Calculation assumes r_{Na^+}, r_{F^-} are hydrated and equal to 1.7 Å, and 2 Å, respectively.

According to the model, a should be a characteristic distance, fixed for each anion–cation pair. For electrolytes whose ionic volumes are large compared to the solvent molecule, e.g., tetrabutylammonium tetraphenyl-boride in acetonitrile, good correlation is found between the sum of their crystallographic radii and a. For these systems the conductivity equation becomes a two-parameter function of Λ^∞ and K_Λ. Furthermore, a reasonable correlation is observed between a_R, a_Λ, and a_K, where a_R is derived from (2.23), a_Λ from Stokes law and the limiting ionic mobility, a_K from the association constant for pairs of ions in contact. For systems where the ionic volume is comparable to the volume of a solvent molecule, it has been found that (i) for a given salt, a increases systematically as the relative permittivity of the solvent decreases, e.g., CsI in dioxane–water mixtures; (ii) for a given solvent, a is independent of the salt, from alkali halides to tetrabutylammonium tetraphenyboron. These results clearly illustrate the failure of the primitive model and have led to the proposal that a should be identified with Bjerrum's critical distance, q, equation (2.27). q represents the radius of the spherical surface around an ion inside of which the mutual attraction of two ions becomes predominant over their kinetic energy. Since, in this case, q would be an inverse function of ε, it would explain the observed behavior of a and its dependence on the solvent rather than the ionic radii. The Bjerrum concept has been carried even further by Justice who equates K_Λ with the Bjerrum association constant[20]

$$K_\Lambda = \frac{4\pi N_A}{1000} \int_a^q r^2 \exp\left(\frac{z^2 e^2}{r\varepsilon kT}\right) dr \qquad (2.28)$$

rather than

$$K_\Lambda = \frac{4\pi N_A a^3}{3000} \exp\left(\frac{z^2 e^2}{a\varepsilon kT}\right) \qquad (2.29)$$

which was derived by Fuoss for ions in contact.[21] For K_Λ values obtained from conductance data, $\ln K_\Lambda$ versus $1/\varepsilon$ plots should be linear according to (2.29); such plots, however, are observed to exhibit curvature at higher values of ε, which is predicted by equation (2.28).[1]

Justice and colleagues[20] have determined the best fit values of Λ^∞, K_Λ, and J_2 to equation (2.23) for a wide range of systems by setting $a = q$ in (2.24) and in the J_1 term of (2.23). The standard deviation in Λ obtained by this method is small, in the range from 0.004 to 0.2%. However, Fuoss[22] has suggested that the form of equation (2.23) (which is based on the work of Fuoss and Hsia) plus the fact that it contains at least two adjustable parameters, Λ^∞ and K_Λ, guarantees a good fit. Furthermore, if (2.23) is regarded as a three-parameter equation, Λ^∞, K_Λ, and a, then a range of values could be found that would give an equally good fit. This is apparent from the work of Beronius and Hanna *et al.* Beronius[23] has shown that for a number of systems there are two possible choices of a that minimize the deviation of (2.23) from the experimental data: one considerably different from q and, for some systems, one where $a \doteq q$, Table 2-1. Hanna *et al.*,[16] using a similar set of equations to Justice and colleagues, found that for $MgSO_4$, $ZnSO_4$, $CuSO_4$, and $MnSO_4$ the best-fit value of a occurs at 10. 7 Å instead of 14.3 Å as predicted from (2.27). Very recently Carman[10] has obtained the most comprehensive solution to the relaxation field and derived a very accurate conductivity equation. He suggests that the most probable values of K_Λ are those corresponding to values of a that agree closely with the estimates from the ionic radii. This criterion, however, can only be maintained for Li^+-, Na^+-, F^--containing salts if these ions are assumed to be hydrated with a radius of 1.5 Å (Table 2-1).

The evidence strongly indicates that the primitive model is inadequate, and even the substitution of the Bjerrum concept of ion pair formation does not remove the basic fault from the theory. The continuum model neglects all short-range ion–ion, ion–solvent interactions and assumes that the bulk properties of the solvent are unaltered in the presence of an ion. Acknowledging this as the fundamental fault of the model, Fuoss said in an address to the 1964 International Symposium of the Electrochemical Society in Toronto,[4c]

> The current model is a good first approximation, but it must now be improved. On continuum theory specific ion solvent interaction is ruled out by the definition of a continuum. Evidently the next step must be to include in the theory the discrete structure of the solvent, at least through the first few nearest neighbor layers.

Fuoss,[11] however, apparently decided against this line of approach since his latest model presents a treatment of long-range electrostatic and hydrodynamic interactions based on general boundary conditions that are independent of short-range forces. The conditions he uses are electroneutrality, continuity of the potential and field gradient at $r = R$, that an unpaired ion contains no other ion within a radius of $r = R$, and that the asymmetry in the distribution function vanishes for ions at infinite separation. R is a parameter defined as the distance from a reference ion beyond which continuum theory may be applied. Each ion is therefore surrounded by a sphere of radius R, outside of which the properties of the solvent are unaltered. Inside the sphere the field strength of the ion (approximately 10^6 V cm^{-1}) polarizes the solvent structure such that its relative permittivity and viscosity will be significantly different from the bulk value. Such spheres are termed "Gurney cospheres." Ions that approach to within $r \leqslant R$ are counted as ion pairs; however, two kinds of ion pairs are identified: solvent separated pairs $a < r \leqslant R$ and contact pairs at $r = a$. Contact ion pairs react to the external field like a real dipole and contribute only to the charging current. Solvent separated pairs are felt as virtual dipoles by unpaired ions and are not included in the calculation of the long-range interionic effects. They do contributed to the conductance current because solvent separated ions execute normal Brownian motion which, under the influence of an external field, is biased in the field direction.

If the concentration of unpaired ions is c_γ and if α is the fraction of paired ions present as contact ion pairs, then the equilibrium constant K_R for the process of solvent separated ion pair $(A^+ \cdots B^-)$ formation is

$$A^+ + B^- \rightleftharpoons (A^+ \cdots B^-) \rightleftharpoons A^+ B^- \ (\rightleftharpoons AB)$$

$$K_R = (1 - \alpha)(1 - \gamma)/c_\gamma^2 f_\pm^2 \tag{2.30}$$

where $\ln f_\pm = -\beta\kappa/2(1 + \kappa R)$.

For the formation of contact ion pairs $(A^+ B^-)$ the equilibrium constant is

$$K_s = \alpha/(1 - \alpha) = \exp(-E_s/kT) = e^{-\varepsilon}$$

where E_S is the energy difference between a pair in the states of $r = R$ and $r = a$, and, therefore,

$$K_\Lambda = (1 - \gamma)/c_\gamma^2 f^2$$
$$= K_R/(1 - \alpha) = K_R(1 + K_s)$$

By assuming that the rate of formation and decomposition of $(A^+ \cdots B^-)$ ion pairs is diffusion controlled, Fuoss calculated the formation κ_1 and decomposition κ_2, rate constants, and hence K_R where

$$K_R = \frac{\kappa_1}{\kappa_2} = \frac{4\pi N_A R^3}{3000} \exp\left(\frac{\beta}{R}\right)$$

The conductivity equation now becomes

$$\Lambda = [1 - \alpha(1 - \gamma)]\left[\Lambda\left(1 + \frac{\Delta E}{E}\right) + \Delta\Lambda_e\right] \tag{2.31}$$

where the first term in brackets is the fraction of ions contributing to the conductance, $\Delta E/E$ is the total relaxation term equation (2.13) including

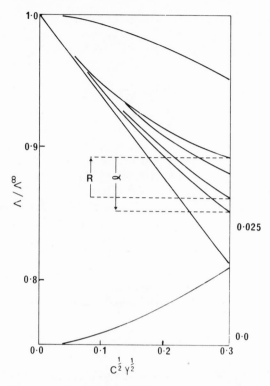

FIG. 2-4. Composition of conductance curve for CsBr in water at 25°C. From top to bottom: χ, Λ_i/Λ^∞, $\Lambda_{LL}/\Lambda^\infty$, $\Lambda_{L,ln}/\Lambda^\infty$, $\Lambda_{(c)}/\Lambda^\infty$; limiting tangent, ordinate scale at left; $(\Lambda_i - \Lambda_{L,ln})/\Lambda^\infty$, ordinate scale at right. Reprinted with permission from R. M. Fuoss, *J. Phys. Chem.* **82**, 2427–2440 (Fig. 2) (1978). Copyright 1978 American Chemical Society.

$\Delta E_v/E$, $\Delta\Lambda_e$ is the term arising from the electrophoretic velocity v_e, equation (2.19).

Figure 2-4 illustrates the fractional contribution made to the total conductivity by the various terms of $\Delta E/E$ and $\Delta\Lambda_e$. The limiting tangent in Fig. 2-4 is given by

$$\frac{\Lambda_{LT}}{\Lambda^\infty} = 1 - \left(\alpha_0 + \frac{\beta_0}{\Lambda^\infty}\right)c^{1/2}\gamma^{1/2}$$

the leading terms of $\Delta E_E/E$ and v_e combine to give

$$\frac{\Lambda_{LL}}{\Lambda^\infty} = 1 + \frac{\Delta E_0}{E} - \frac{\beta_0 c^{1/2}\gamma^{1/2}}{1+t}$$

and the terms to $c \ln c$ give

$$\frac{\Lambda_{L,\ln}}{\Lambda^\infty} = 1 + \frac{\Delta E_0}{E} + \frac{\tau^2}{3} \ln t - \beta_0 c^{1/2} \gamma^{1/2} \left(\frac{1}{1+t} + \frac{\tau}{4} \ln t \right)$$

When all the terms in $\Delta E/E$ and v_e are included, the result is graphed as

$$\frac{\Lambda_i}{\Lambda^\infty} = 1 + \frac{\Delta E}{E} + \frac{\Delta \Lambda_e}{\Lambda^\infty}$$

which is the conductivity curve when there is no ion pair formation. The curve $\Lambda(c)/\Lambda^\infty$ is obtained by multiplying Λ_i/Λ^∞ by $\chi = [1 - \alpha(1 - \gamma)]$. The magnitude of R is related to the difference between the Λ_i/Λ^∞ and $\Lambda_{L,\ln}/\Lambda^\infty$, as shown in Fig. 2-4, and α to the difference between Λ_i/Λ^∞ and $\Lambda(c)/\Lambda^\infty$. Increasing R has the effect of increasing Λ_i/Λ^∞ and hence $\Lambda(c)/\Lambda^\infty$, while increasing α moves $\Lambda(c)/\Lambda^\infty$ downward toward the limiting tangent. These two opposing effects often result in a range of paired α and R values that fit the calculated conductivities to the experimental data within the limits of experimental precision.

As with previous conductance equations, equation (2.31) is a three-parameter equation

$$\Lambda = \Lambda(\Lambda^\infty, R, \alpha)$$

and the most unambiguous values of R and α are determined for solutions of high-dielectric constant, $\varepsilon > 25$. For example, for LiCl in a water–dioxane mixture of $\varepsilon = 62.25$, R can vary by $\pm 10\%$ from the best fit value of 7.09 Å, and the calculated Λ will still fit the experimental data to 0.015%. However, for $\varepsilon \leqslant 25$ a progressively wider range of R values can be tolerated for a good fit to the data. As R is varied across a range of best-fit values, the corresponding values of Λ^∞ and K_Λ also vary, the former by very little, but K_Λ becomes increasingly independent of R as the dielectric constant of the solvent is reduced. An explanation for this observation is found in equation (2.30); as R, the diameter of the cosphere, increases with decreasing ε, more ions are counted as paired, thereby reducing γ but at the same time increasing the activity coefficient $f_\pm$. Both effects tend to cancel each other, with the result that K_Λ becomes insensitive to the value of R.

For solutions where $\varepsilon > 25$, Fuoss has arbitrarily preset the center-to-center distance of a solvent-separated ion pair as $R = a + d$, where a is the sum of the ionic radii and d is an average distance

$$d = \left(\frac{M}{N_A} \right)^{1/3}$$

where M is the molecular mass of the solvent, ρ is its density, and N_A is Avogadro's number. While this is a simplistic approach to the determination of R, it provides a self-consistent way of comparing data from different solvents of low relative permittivity. For systems where $a + d < \beta/2 = q$, R is arbitrarily set equal to $\beta/2 = q$, the Bjerrum distance.

This theory is an improvement on the primitive model because long-range coulomb effects are taken to depend only on the charges of the ions and not on their size. The short-range effects, ion–solvent and solvent–solvent, are subsumed into one parameter E_s, and the distance parameter R will be a function of surface charge density and solvent ε.

2.2. The Concentration Dependence of Conductivity at High Pressure and Moderate Temperatures

According to the primitive model of conductivity in electrolyte solutions, the concentration dependence of Λ for a given solute–solvent system at constant temperature and pressure is determined by the dielectric constant, the viscosity of the solvent, and the a parameter. Thus it is not surprising that the equations derived from the continuum model for systems at atmospheric pressure seem to be equally applicable to the same systems at elevated pressures within the studied range of 5 kbar and the experimental precision of 0.1%. The effect of pressure on most common solvents is to alter the relative permittivity and to increase the viscosity. To the same extent to which the continuum model is valid at atmospheric pressure, it should also be valid at higher pressures. However, because of the experimental difficulties involved in precise high-pressure measurements it has not been possible to test rigorously the conductivity equations as has been done at atmospheric pressure. Furthermore, most studies have been designed to determine the gross effects of pressure on Λ^∞ and as such have been concerned with the validity of techniques used in extrapolating data to infinite dilution.

The applicability of the Onsager limiting law in extrapolating data from higher concentrations has been reviewed by Hamann[24] and more comprehensively by Brummer and Gancy.[25] The latter authors refer mostly to their own painstaking work[26]; they have measured the effects of pressure on the conductivity of eleven 1 : 1 electrolytes in water up to 2300 atm and 55°C to a precision of 0.05%. For the purpose of analyzing their data, they prefer to consider the ratio Λ_p/Λ_1, whose concentration dependence can be written as

$$\frac{\Lambda_p}{\Lambda_1} = \frac{\Lambda_p^\infty - F(c_p)}{\Lambda_1^\infty - F(c_1)} \tag{2.32}$$

where $F(c_p)$ and $F(c_1)$ are appropriate functions of the concentration at pressure p, and at 1 atm, respectively. If these functions are defined by the Onsager limiting law, then it can be shown that

$$\frac{\kappa_p}{\kappa_1} = \left(\frac{\kappa_p}{\kappa_1}\right)_{c \to 0} \left\{ 1 + \left[\frac{S_1}{\Lambda_1^\infty} - \frac{S_p}{\Lambda_p^\infty} \left(\frac{\rho_p}{\rho_1}\right)^{1/2} \right] (c_1)^{1/2} \right\} + \text{higher terms in } (c_1)^{n/2}$$

(2.33)

where $(\kappa_p/\kappa_1)_{c \to 0}$ is the value at infinite dilution and S is given by equation (2.1). Ignoring the higher terms, equation (2.33) predicts a linear dependence of κ_p/κ_1 on $(c_1)^{1/2}$. This is in fact observed for LiCl, NaCl, KCl, RbCl, and NH_4Cl at 1000 and 2000 atm and 25°C from 3×10^{-3} mol dm^{-3} to 20×10^{-3} mol dm^{-3} to within ±0.1%. Obviously this is well beyond the range of validity of the limiting law and is due to the cancelation of deviations from the law in the conductivity ratio. Nakahara and others[27] have observed the same result for KCl to 5 kbar and have shown that the ratio $(\kappa_p/\kappa_1)/(\kappa_p/\kappa_1)_{c \to 0}$, as given by equation (2.33), is greater than unity for a 10^{-2}-mol-dm^{-3} solution and increases with pressure but decreases with increasing temperature at constant pressure. These trends are reversed in very concentrated solutions; see Chapter 3.

At higher concentrations, (κ_p/κ_1) exhibits a significant departure from the $(c_1)^{1/2}$ dependence; a more sophisticated $F(c_1)$ function is required. Instead of using the more rigorously derived equation due to Fuoss and Hsia, Brummer and Gancy chose the more convenient expression of Robinson and Stokes[28]

$$F(c_p) = \frac{S_p(c_p)^{1/2}}{1 + \kappa_p a_p}$$

(2.34)

This is merely (2.1) divided by $(1 + \kappa a)$ to allow for finite ion size; this equation fits 1-atm conductivity data for 1 : 1 electrolytes to about 0.05% up to 0.05 mol dm^{-3}, equivalent to the experimental precision pertaining in high-pressure work. When substituted into (2.32), and after dropping all terms of higher power than c, the result is

$$\frac{\kappa_p}{\kappa_1} = \left(\frac{\kappa_p}{\kappa_1}\right)_{c \to 0} \left\{ 1 + \left[\frac{S_1}{\Lambda_1^\infty} - \frac{S_p}{\Lambda_p^\infty} \left(\frac{\rho_p}{\rho_1}\right)^{1/2} \right] (c_1)^{1/2} \right.$$

$$\left. + \frac{S_1}{\Lambda_1^\infty} \left[\frac{S_1}{\Lambda_1^\infty} - \frac{S_p}{\Lambda_p^\infty} \left(\frac{\rho_p}{\rho_1}\right)^{1/2} \right] c_1 + \kappa_1 a_1 \frac{a_p}{a_1} \left(\frac{\varepsilon_1}{\varepsilon_p}\right)^{1/2} \left(\frac{\rho_p}{\rho_1}\right) - \frac{S_1}{\Lambda_1^\infty} \right\} c_1$$

(2.35)

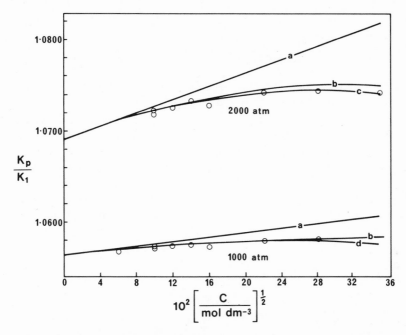

FIG. 2-5. Higher concentration dependence of the pressure coefficient of conductance of NaCl solutions at 25°C; a, limiting tangent; b, equation (2.35) with $a_1 = 6.1$ Å and $a_p/a_1 = 1.0$; c, equation (2.35) with $a_1 = 4.4$ Å and $a_p/a_1 = 0.975$; d, equation (2.35) with $a_1 = 4.4$ Å and $a_p/a_1 = 0.994$. Reprinted with permission from A. B. Gancy and S. B. Brummer, *J. Phys. Chem.* **73**, 2429–2436 (Fig. 6) (1969). Copyright 1969 American Chemical Society.

For aqueous NaCl it can be seen in Fig. 2-5 that this equation fits the data to 0.01 mol dm^{-3} at 1000 and 2000 atm depending on the choice of a_1 and a_p/a_1. If it is assumed that $a_p/a_1 = 1$ at all pressures (i.e., that a is the sum of the radii of incompressible ions), then a must fall within the range $a = 6.0$–6.4 Å for equation (2.35) to fit the data (Fig. 2-5, curve b). In fact, the chosen value of $a_1 = 6.1$ Å was determined by Fuoss and Hsia as providing the best fits to their equation for slightly associated NaCl in water, 8% at 0.01 mol dm^{-3}. Since (2.35) takes no account of ion pairing, the good fit at high concentrations must arise again from cancelations of deviations in the conductivity ratio.

Curves c and d in Fig. 2-5 were calculated by setting a_1 equal to the value obtained from conductivity data for NaCl and by adjusting the ratio a_p/a_1 to obtain the best fit.

Another approach to extrapolating conductance data to infinite dilution, in solvents of low dielectric constant where ion association occurs, has been to use the Shedlovsky equation[29]

$$\frac{1}{\Lambda F(z)} = \frac{1}{\Lambda^\infty} + \frac{c\Lambda f_\pm^2 F(z)}{K_\Lambda \Lambda^{\infty 2}} \qquad (2.36)$$

This equation is based on (2.1) and $z = (S/\Lambda^{\infty 3/2})(c\Lambda)^{1/2}$, a plot of $1/\Lambda F(z)$ versus $c\Lambda f_\pm^2 F(z)$ determines Λ^∞ and K_A.

It has been shown to fit the data within experimental error for $(CH_3)_4NBr$ and HBr in propan-1-ol and propan-2-ol,[30] and for $(CH_3)_4NI$, $(CH_3CH_2)_4NI$, $(CH_3CH_2CH_2)_4NI$ in acetone[31] at high pressures. Because of ion pairing none of these systems except HBr gave limiting slopes that agreed with the theoretical value. The more sophisticated conductivity expression [equation (2.20) where $J_2 = 0$] due to Fuoss and Onsager has been tested for several salts in methanol[32] and for NaI,

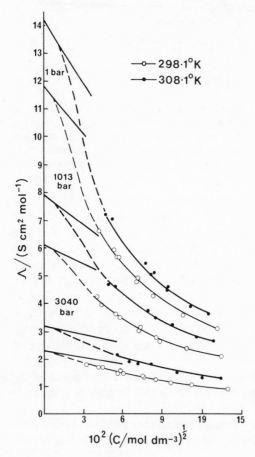

FIG. 2-6. Conductivity of NaI in 2 methylpropan-1-ol as a function of concentration at 1, 1013, and 3040 bar. The solid curves are calculated from equation (2.25). Reprinted with permission from A. H. Ewald and J. A. Scudder, *Aust. J. Chem.* **23**, 1939–1945 (Fig. 1) (1970).

KI, and CsI in water.[33] In the latter case the authors were more concerned with extracting value for Λ^∞ and do not comment on the fit of the data to the equation. However, in the former case Fuoss and Skinner relate the degree of ion pairing to the magnitude of the *a* parameter required to obtain the best-data fit. Unfortunately their data are not sufficiently precise to warrant further examination by equation (2.25) and prefer instead an approximation to it derived by Fuoss and Kraus,[34] which allows for ion association. Equation (2.25), however, has been tested for several salts in acetone and 2-methylpropan-1-ol.[35] Figure 2-6 displays results for NaI at various pressures; for an arbitrary value of $a = 5$ Å the data seem to fit equation (2.25) within experimental error. These curves are described by Fuoss and Accascina as Type III phoreogram[5]; the diminishing effect of increasing concentration on Λ is due to an increase in the degree of association. The data for Bu_4NI in propan-2-ol are more precise and show that the fit to (2.25) becomes worse at higher pressures because of accumulated experimental errors.[36]

In summary, the primitive continuum model has been substantiated as a useful description of electrolyte behavior at high pressures.

2.3. The Concentration Dependence of Conductivity at High Pressure and Temperature

The study of conductivity in electrolyte solutions at high temperatures and pressures is of interest to geochemists, industrial chemists, corrosion engineers, and many other applied scientists. Consequently, whilst many studies have been made of the phenomenon, relatively few have been concerned with a quantitative physiocochemical interpretation of experimental data. Gancy and Brummer have reviewed the field up to 1972 and have provided an extensive tabulation of papers, along with a discussion of experimental techniques and methods of data analysis.[25] There have been few developments in this area since then, and this section will necessarily review some of that material.

Two schools of research, those of E. U. Franck at the Institut Für Physikalische Chemie und Elektrochemie der Universitat Karlsruhe and of S. Quist and W. L. Marshall of Oak Ridge National Laboratory, Tennessee, have provided the most suitable compilation of data for analysis in terms of classical electrolyte theory. Much of their work was conducted on solutions under experimental conditions that extend well beyond the supercritical temperature and pressure of water and, in dilute solutions, of the solution itself. However, a continuous transition of bulk solvent properties, relative permittivity, viscosity, and solution conductance is observed through the critical point, and there is no *a priori* reason to suppose that the primitive continuum model is inapplicable to these systems.

A typical example of the data from Quist and Marshall's numerous publications is their study of aqueous NaCl solutions from 0.001 to 0.1 mol dm^{-3} over a range of 800°C and 4 kbar, $\rho = 0.2$ g cm^{-3} to 1.1 g cm^{-3}.[37] Above $\rho = 0.75$ g cm^{-3}, where ion association is low, equations (2.34) and (2.20) (where $J_2 = 0$) represent the data within the limits of experimental accuracy $\sim 1\%$. For (2.34) a value of $a = 2\text{Å}$ and for (2.20) $a = 3-4$ Å provided the best fit; both values are within reasonable limits of that required by the continuum model. Below $\rho = 0.65$ g cm^{-3} neither of these equations fits the data. This is not unexpected since the dielectric constant of water drops significantly with increasing temperature and, by equations (2.28) and (2.29), ion pairing becomes more significant. The Shedlovsky function (2.36) then becomes the most suitable method for obtaining Λ^{∞} and K_Λ. The values of Λ^{∞} and $1/K_\Lambda$ obtained from the above procedures are displayed in Figs. 2-7 and 2-8. K_Λ behaves qualitatively in the manner that would be predicted on the basis of equations (2.28) and (2.29); at constant temperature a decrease in density causes a decrease in ε for water and hence K_Λ the association constant increases. Similar results have been obtained for aqueous solutions of K_2SO_4, H_2SO_4, $KHSO_4$, NaBr, NaI, HBr, NH_4OH, NaOH.[38] Franck and co-workers have studied the electrical conductivity of aqueous solutions of KCl,[40] LiCl, CsCl,[40] $BaCl_2$, $Ba(OH)_2$, and $MgSO_4$[41] over a wide range of temperatures and pressures. Their studies on KCl solutions are the most

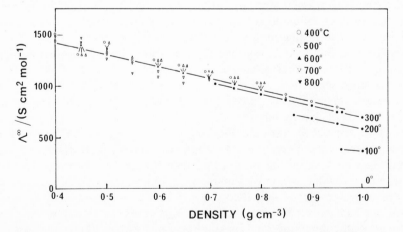

FIG. 2-7. Λ^{∞} for NaCl as a function of density. Reprinted with permission from A. S. Quist and W. L. Marshall, *J. Phys. Chem.* **72**, 684–703 (Fig. 24) (1968). Copyright 1968 American Chemical Society.

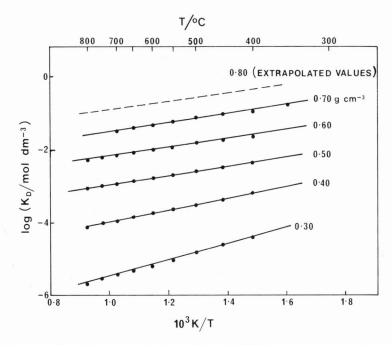

FIG. 2-8. Log K_D, the dissociation constant for $NaCl = Na^+ (aq) + Cl^- (aq)$ versus $1/T$ at several densities. Reprinted with permission from A. S. Quist and W. L. Marshall, *J. Phys. Chem.* **72**, 684–703 (Fig. 28) (1968). Copyright 1968 American Chemical Society.

comprehensive and in three papers span the temperature, pressure, and density range of 299–1000°C, 1–12,000 bar, and 0.2–1.2 g cm^{-3}. The first paper covers the range of 750°C, 2800 bar, and 0.2–1.0 g cm^{-3}.[39] At densities greater than 0.7 g cm^{-3} plots of Λ versus $c^{1/2}$ are linear up to 0.01 mol dm^{-3} (Fig. 2-9) in agreement with equation (2.1). However, below these densities ion association becomes significant and the plots are nonlinear. The Shedlovsky function, equation (2.36), proved to be a successful method of obtaining Λ^∞ and K_Λ up to 500°C and densities greater than 0.6 g cm^{-3}. Below this density the data were not sufficient for an accurate extrapolation, and the values of Λ^∞ were obtained by assuming the applicability of the Walden rule. Hence $\Lambda(T, \rho)$ was obtained from the known $\Lambda(400°C, 0.8 \text{ g cm}^{-3})$, $\eta(400, 0.8)$, and $\eta(T, \rho)$ via

$$\Lambda^\infty(T, \rho) = \frac{\Lambda(400, 0.8)\eta(400, 0.8)}{\eta(T, \rho)}$$

The second study covered the same temperature range but increased the total pressure to 6 kbar and covered the density range from 0.5 to 1

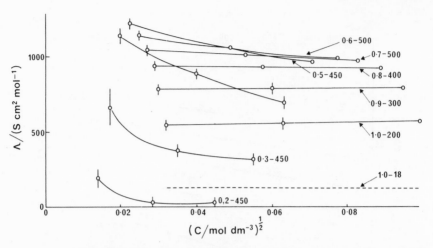

FIG. 2-9. Λ for aqueous KCl solutions as a function of $c^{1/2}$, at various densities in g cm^{-3}, and temperatures in degree celsius. Reprinted with permission from E. U. Franck, *Z. Phys. Chem.* **8**, 92–126 (Fig. 4) (1956).

g cm^{-3}.[41] The concentrations range from 0.01 to 0.001 mol kg^{-1} of KCl was sufficient, at these densities, to allow an extrapolation by the Fuoss and Kraus equation to obtain Λ^{∞} and K_{Λ}. The final study extended the pressure to 12 kbar, 1000°C and covered the density range 0.7 to 1.2 g cm^{-3}.[40] Limiting molar conductivities were calculated from the Walden rule using water viscosities extrapolated from lower temperatures, and these were substituted into the Shedlovsky equation to obtain κ and α, the degree of dissociation where $\alpha = (\Lambda/\Lambda^{\infty})F(z)$. Some of the combined results for Λ^{∞} from three studies are shown in Fig. 2-10. As for NaCl, Λ^{∞} seems to become relatively insensitive to temperature above 300°C. The dissociation constant for KCl also follows a similar pattern to that observed for NaCl. For 0.01-mol-dm^{-3} KCl, α varies between 0.75 at $\rho = 0.70$ over 300 to 1000°C to 0.96 at $\rho = 1.10$ and 300°C; the implication is that KCl is not fully dissociated above 300°C even at very high pressures. However, in view of the recent results obtained for KCl at 25°C and 1 atm pressure this is not surprising (see Table 2-1).

In 1956 Franck[39] introduced the concept of the complete ionization constant k^0; this concept was further pursued by Quist and Marshall,[38] who, for the dissociation of NaCl, wrote

$$\text{NaCl(aq)} + h\text{H}_2\text{O} \rightleftharpoons \text{Na}^+ \text{(aq)} + \text{Cl}^- \text{(aq)}$$

where

$$k° = \frac{a_{\text{Na}^+(\text{aq})} a_{\text{Cl}^-(\text{aq})}}{a_{\text{NaCl}(\text{aq})} a_{\text{H}_2\text{O}}^h}$$

Thus $k°$ differs from the conventional ionization constant by the inclusion of the water activity. The standard state for the solute was chosen as the hypothetical 1 M solution whose thermodynamic properties are the same as a solution at infinite dilution at the particular pressure, and the activity of water was regarded as being equal to its concentration. Thus $\log k = \log k° + h \log C_{\text{H}_2\text{O}}$, and a plot of $\log k$ versus $C_{\text{H}_2\text{O}}$ should be linear with a slope equal to h, the hydration number. Furthermore, it was argued that $k°$ should be independent of solvent dielectric constant and pressure and dependent only on temperature. Such behavior has been observed for a number of electrolytes in water and for NaCl in water–dioxane mixtures.[43] However, this concept has been criticized on two grounds. Gilkerson[44] has shown from experimental data that the $a_{\text{H}_2\text{O}}$ is more pressure dependent than $C_{\text{H}_2\text{O}}$ and therefore $k°$ is pressure dependent. From a fundamental thermodynamic basis Matheson[45] has shown that there is no thermodynamic reason why $k°$ should necessarily be independent of pressure, and therefore he rejects the concept as defined.

Within the limits of the substantial experimental error involved in this difficult field of research, it would appear that the continuum model of

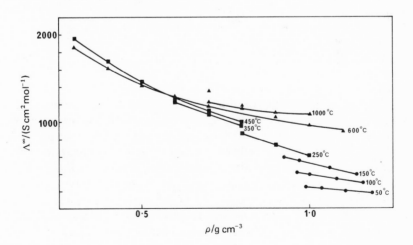

FIG. 2-10. Λ^∞ for aqueous KCl as a function of density at various temperatures; data from Refs. 39, 41 and 42.

electrolyte solutions developed for systems at ambient temperatures and pressures is applicable to aqueous and mixed solvent systems well into the supercritical region at low and high densities.

2.4. The Effect of Pressure on Electrical Conductivity at Ambient Temperatures

There have been numerous studies carried out on electrolyte systems below 100°C and up to a few kbar. Some have been carried out only at one concentration so that Λ^∞ values were not obtainable, and as such they do not provide much direct insight into the effects of pressure on the mobility of ions because of the complicating feature of ion–ion interactions. Fortunately recent investigators have been able to obtain sufficient data to extrapolate to infinite dilution and contribute toward the significant body of data that is building a useful picture of the dynamic effects of ion–solvent interactions.

Most experiments have been carried out with water as solvent. However, the results from studies on nonaqueous systems provide some very important information as these solvents do not exhibit the viscosity anomaly shown by water at high pressures. Furthermore, the solvent can be chosen so that it selectively solvates one or both ions; a range of viscosities and dielectric constants is also available.

The system of solvent classification suggested by Kay, Evans, and Matesich will be adopted here.[46]

2.4.1. Hydrogen-Bonded Solvents

2.4.1.1. Aqueous Solutions

The effect of pressure on the limiting ratio of molar conductivities of several electrolytes is shown in Fig. 2-11. At ambient temperatures, up to 25°C, the ratio increases with pressure to a maximum in the vicinity of 1 to 1.5 kbar, after which it continues to decline. Figure 2-12 illustrates that the increase in $(\lambda_p/\lambda_1)^\infty$ is due almost solely to the Cl^- ion, which exhibits a positive pressure dependence up to 1.5 kbar at least. However, λ^∞ for the alkali metal ions also increases on the initial application of pressure. This phenomenon, which is not observed at or above 40°C, has been attributed to the breakdown of water structure with increasing pressure as reflected by the viscosity that passes through a minimum at 1 kbar[48], the minimum disappearing above 30°C; Fig. 2-13. This cannot be the sole explanation, however, since λ^∞ for many ions does not exhibit a maximum, and most

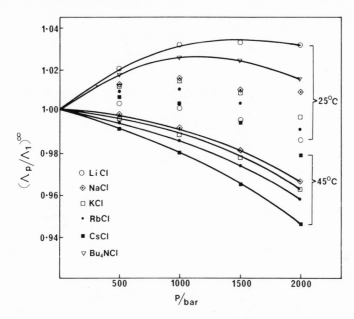

FIG. 2-11. Λ_p/Λ_1 versus pressure for alkali metal chloride salts and Bu_4NCl in water at 25 and 45°C.[26a]

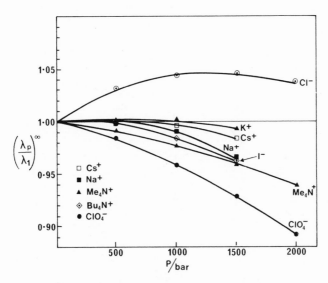

FIG. 2-12. $(\lambda_p/\lambda_1)^\infty$ versus pressure for various ions in aqueous solution at infinite dilution and at 25°C. Data from Refs. 33, 47.

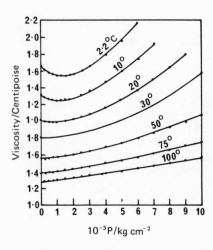

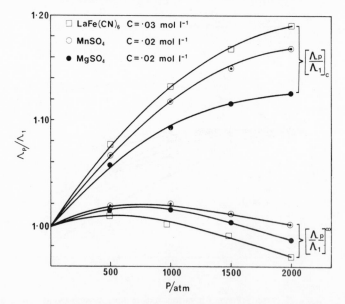

FIG. 2-13. Viscosity of water versus pressure at different temperatures. Reprinted with permission from K. E. Bett and J. B. Cappi, *Nature* **207**, 620 (Fig. 2) (1965).

ions show significant departures from Walden's rule with increasing pressure, Section 2.7.

For LiCl, KCl, and NaCl increasing concentration appears to have little influence on the shape of the curves in Fig. 2-11 up to at least 1.0 molal,[49, 26b] and this is probably true for the remaining alkali halides. However, for $(CH_3)_4NCl$ and Bu_4NCl a comparison of two sets of data

FIG. 2-14. Conductivity ratios for some associated electrolytes versus pressure at finite concentrations and at infinite dilution. Data from Refs. 52, 53, and 56.

indicates that $\Lambda_p/\Lambda_1^{(50)}$ for a 0.1 M solution is less than $(\Lambda_p/\Lambda_1)^{\infty(51)}$ for all pressures, and from the data in reference (49) the ratio for a 1 molal solution of Me_4NCl is even lower. This is the usual trend in concentrated solutions, however, it occurs at lower concentrations for the tetraalkylammonium salts.

Associated electrolytes at finite concentrations exhibit quite different conductivity curves to those shown at infinite dilution, Fig. 2-14. Λ^∞ for $MgSO_4$ at a given temperature and pressure was calculated from the equivalent data for $MgCl_2$, K_2SO_4, KCl, and Kolhrauch's law of the independent mobility of ions.[52] If the further assumption is made that the mobility of an ion at finite concentration can be calculated by classical electrolyte theory, then the ratio of $\Lambda_{measured}/\Lambda_{theory} = \gamma$, the degree of dissociation. From this and the estimated activity coefficients, the molal dissociation constant, K_m, was calculated. The data for the degree of association $(1 - \gamma)$ for $MgSO_4$ are displayed in Table 2-2. Clearly $MgSO_4$ becomes more ionized with increasing pressure, and the increase in the concentration of ions counteracts to some extent the effect of pressure on the mobility of the ion, and the conductivity maximum is shifted to higher pressures with increasing concentration. A similar analysis has been made of the data for $MnSO_4$,[53] K_2SO_4, Na_2SO_4, $MgCl_2$,[54] Li_2SO_4, Rb_2SO_4, Cs_2SO_4, and $(NH_4)_2SO_4$.[55]

The partial molal volume change ΔV_m^∞, defined as

$$\Delta V_m^\infty = -RT\left(\frac{\partial \ln K_m}{\partial p}\right)_{T,m} \tag{2.37}$$

calculated from the experimental data for $MgSO_4$ and $La_3Fe(CN)_6$ is displayed in Table 2-3, along with the best fit values of a to equations (2.28) and (2.29) and the calculated values of ΔV_m^∞ obtained from differentiating equations (2.28) and (2.29).[53] Both equations adequately account for the effect of pressure on K_m within the large experimental errors

TABLE 2-2. Degree of Association $(1-\gamma)$ for Aqueous $MgSO_4$ at $25°C^a$

	P/atm				
$10^4 c/mol\,dm^{-3}$	1	500	1000	1500	2000
5.00	0.067	0.059	0.056	0.043	0.037
10.01	0.107	0.096	0.089	0.075	0.068
20.00	0.158	0.144	0.133	0.115	0.105
100.1	0.314	0.290	0.271	0.248	0.230
200.1	0.386	0.360	0.340	0.315	0.297

TABLE 2-3. ΔV_m^∞ Calculated from Equations (2.28) and (2.29) for Ion-Pair Dissociation at 25°C and 1 atm[a]

	MgSO₄	LaFe(CN)₆
$\Delta V_m^\infty / cm^3 \, mol^{-1}$		
Equation (2.28)	− 4.86	− 6.89
Equations (2.29)	− 7.42	− 8.98
Experimental	− 7.30	− 8.00
Best fit $a/Å$		
Equation (2.28)	4.19	7.04
Equation (2.29)	3.87	7.36

[a]Data from Ref. 56.

associated with this type of calculation. Note that ΔV_m^∞ is negative in all cases indicating a volume decrease on formation of ions. This decrease is believed to arise from the electrostriction of water due to electrostatic interaction between ions and water molecules.

The high-pressure research school of Kyoto University has recently published a number of papers concerned with the effects of pressure on ionic mobility and ion association in aqueous solutions at high pressures.[57] Some of their results are displayed in Fig. 2-15 and exhibit a temperature and pressure dependence on the dielectric constant that would be predicted from equations (2.28) and (2.29). Note that for the $Co(NH_3)_6^{2+}SO_4^{2-}$ ion pair, ΔV_m^∞ tends to zero at 5 kbar and may even become positive at 40°C.

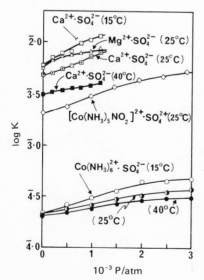

FIG. 2-15. Log K_D, the dissociation constant, for ion pairs in water versus pressure. Reprinted with permission from J. Osugi, K. Shimizu, M. Nakahara, E. Hirayama, Y. Matsubara, and M. Ueno, Proc. 4th International Conference on High Pressure, 610–614 (Fig. 1) (1974).

Electrical conductivity in water–dioxan mixtures at high pressure displays an expected trend away from the anomalous behavior shown by aqueous solutions.[58] Dioxan is a nonhydrogen-bonding solvent, and as the dioxan content increases it is expected that a composition will be reached at which the viscosity anomaly exhibited by pure water will disappear; above this composition Λ^∞ will decrease with increasing pressure. This behavior has indeed been observed at 25 wt% dioxan and above.

2.4.1.2. Alcohols

The alcohols represent an homologous series of solvents of relatively high dielectric constant, whose viscosity increases with molecular mass. Quite a number of electrolytes have been studied in methanol,[59, 60] ethanol,[61] and a few in propan-1-ol, propan-2-ol,[30] and 2 methylpropan-1-ol.[35] Λ^∞ decreases with increasing pressure although the Walden product shows that this is not entirely attributable to viscosity changes, Fig. 2-16. The relative magnitudes of the Walden product, and their pressure dependence, depend in a complex manner on the solute–solvent interactions and are discussed in Section 2.7.

Most of the solutes in Fig. 2-16 are associated to some extent in alcohols. Contrary to expectations, association for NaBr in methanol

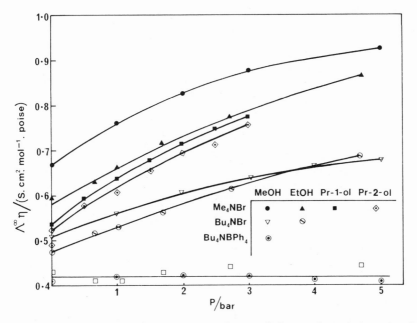

FIG. 2-16. Walden product $\Lambda^\infty\eta$ versus pressure for tetraalkyl ammonium salts in methanol, ethanol, propan-1-ol, propan-2-ol at 25°C. Data from Refs. 30, 60, 61.

increases with increasing pressure, which has been rationalized by assuming that free ions are solvated to a greater extent than the ion pairs and hence a reduction in volume occurs on ion pair of formation. This argument is used to explain why K_Λ decreases with pressures in aqueous solutions. NaBr is apparently unassociated in ethanol, although K_Λ for the larger tetraamyl ammonium salts decreases with increasing pressure. Ion association is quite significant in the higher alcohols, and K_Λ decreases with increasing pressure and on going from propan-2-ol to propan-1-ol, i.e., K_Λ increases with decreasing dielectric constant.

2.4.2. Neutral Solvents

These solvents do not appear to interact strongly with either cations or anions and contrast directly with the solvating properties of hydrogen bonding solvents.

Adams and Laidler[31] made an extensive study of the conductance of Me_4NI, Et_4NI, and Pr_4NI in acetone from 1 to 10^{-4} $mol\,l^{-1}$ from 25 to 55°C and up to 1 kbar. Their data show that the activation volumes $\Delta^{\ddagger} V_i^\infty$ (Section 2.7.2) at a given temperature and pressure are almost independent of the cationic radius, and from 26.61 to 54.83°C range from approximately 11 $cm^3\,mol^{-1}$ at 1 bar to 7 $cm^3\,mol^{-1}$ at 1103 kbar, i.e., they exhibit very similar pressure coefficients of conductance. The Walden products show positive deviations from Walden's rule, which increase with pressure, indicating positive excess mobilities for these ions in acetone. With increasing concentration, ion pair formation becomes significant. The dissociation constants increase with increasing pressure and decrease with increasing temperature as would be predicted from the dielectric constant. However, the ion size parameters, calculated by fitting the data to Fuoss's equation (2.29) are too small to fit the physical model.

The limiting molar conductivities of Bu_4NPi, Me_4NPi, HPi, KPi, KBr salts in N,N-dimethylformamide $(DMF)^{(62)}$ are considerably less than those for the equivalent tetra-amylammonium iodide in acetone; however, at 25°C their activation volumes are similar. At higher pressures $\Delta^{\ddagger} V_i$ tends toward a limiting low value, Fig. 2-17, in agreement with observations on methanol and nitrobenzene solutions but contrary to the findings in acetone solutions. The effect of ion size is small and also appears to be tending toward a limiting value as ionic size increases. The activation volumes of these salts in nitrobenzene are larger than in DMF and, as shown by the data for Bu_4NPi, Fig. 2-18, tend toward a common low value of 8.5 $cm^3\,mol^{-1}$ at all temperatures.[63] $\Delta^{\ddagger} V_i^\infty$ increases slightly with increasing ionic radius, but the data do not exhibit the same systematic dependence on ion size as they do DMF solutions.

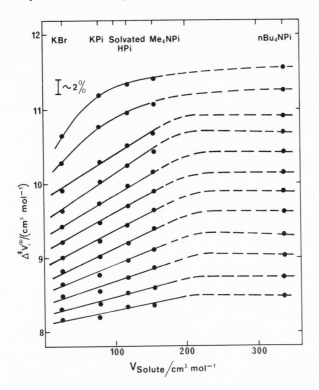

FIG. 2-17. The activation volume $\Delta^{\ddagger}V^{\infty}$ versus molar volume of solute in DMF solvent at 25°C and increasing solvent volumes. Solvent volumes are from bottom to top: 71.5, 72, 72.5, 73, 73.5, 74, 74.5, 75, 75.5, 76, 76.75, and 77.41 cm³ mol⁻¹. Reprinted with permission from S. B. Brummer, *J. Chem. Phys.* **42**, 1636–1646 (Fig. 8) (1965).

For solvents of low dielectric constant, Λ passes through a minimum with increasing concentration.[5] The minimum is due to the onset of triple ion formation and subsequent increase in conductivity. Skinner and Fuoss[64] argue that the effect of pressure on solutions before the minimum will be to increase the conductivity due to increased ion pair dissociation as the dielectric constant increases. However, at concentrations above the conductance minimum increasing pressure should cause a decrease in conductivity as triple ions dissociate to ion pairs. To test this hypothesis they measured the conductance of tetraisoamylammonium picrate (TIAPi) in diethylether and in benzene. The data for TIAPi in diethylether are shown in Fig. 2-19; from the data at 1 atm it is clear that at the highest concentrations Λ is approaching the minimum. At higher pressures the

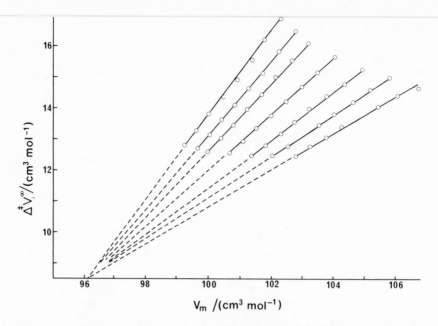

FIG. 2-18. Volumes of activation for conductivity of Bu_4NPi as a function of solvent volume and, from top to bottom, at 20, 25, 30, 40, 50, 60, and 70°C. Reprinted with permission from F. Barreira and G. J. Hills, *Trans. Faraday Soc.* **64**, 1359–1375 (Fig. 7) (1968).

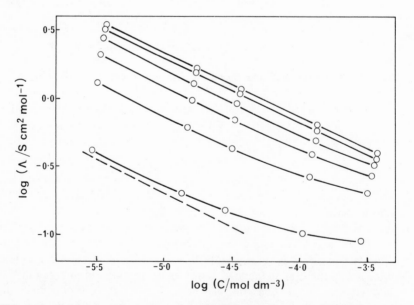

FIG. 2-19. Log Λ versus log c of tetraisoamylammonium picrate in diethylether at 25°C. Bottom to top: 1, 1000, 2000, 3000, 4000, 5000 kg cm^{-2}. Reprinted with permission from J. F. Skinner and R. M. Fuoss, *J. Phys. Chem* **69**, 1437–1443 (Fig. 4) (1965). Copyright 1965 American Chemical Society.

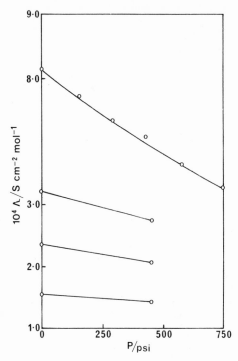

FIG. 2-20. Λ as a function of pressure for tetraisoamylammonium picrate in benzene. Top to bottom: 2.099×10^{-3}, 30°C; 0.9847×10^{-3}, 25°C; 0.6723×10^{-3}, 25°C; and 0.03083×10^{-3}, 25°C. Reprinted with permission from J. F. Skinner and R. M. Fuoss, *J. Phys. Chem.* **69**, 1437–1443 (Fig. 5) (1965). Copyright 1965 American Chemical Society.

conductivity increases and the slope of the $\log \Lambda$ versus $\log C$ plot approaches $-\frac{1}{2}$, the slope predicted when only ions and ion pairs are present. The minimum is pushed to higher concentrations, which become inaccessible because of the solubility limit. However, solutions of TIAPi in benzene can be prepared at concentrations above the conductivity minimum; the effect of pressure on Λ is shown in Fig. 2-20 for four concentrations. Λ decreases with increasing pressure as predicted, the effect becoming more pronounced with increasing concentration. A similar effect was observed for Bu_4NPi in toluene.

2.5. *The Effect of Pressure on the Conductivity of Electrolyte Solutions at High Temperatures and Pressures*

In the absence of interionic effects the conductivity of 1 : 1 electrolytes in aqueous solution appears to become independent of temperature above 400°C but increases linearly with decreasing density (Figs. 2-7, 2-10).

For NaCl, NaBr, NaI, KHSO$_4$, HBr, NaOH, and NH$_4$OH[38] the relation as a function of density, ρ, is described by the linear equations

$$\Lambda^{\infty}(\text{NaCl}) = 1876 - 1160\rho$$

$$\Lambda^{\infty}(\text{NaBr}) = 1880 - 1180\rho$$

$$\Lambda^{\infty}(\text{NaI}) = 1897 - 1210\rho$$

$$\Lambda^{\infty}(\text{KHSO}_4) = 1740 - 1100\rho$$

$$\Lambda^{\infty}(\text{HBr}) = 1840 - 560\rho$$

$$\Lambda(\text{NaOH})^{\dagger} = 1770 - 630\rho$$

$$\Lambda(\text{NH}_4\text{OH})^{\dagger} = 1910 - 630\rho$$

The data for KCl are probably less precise than those of the aforementioned studies but cover a greater temperature and density range. Within experimental error, however, they appear to extrapolate to a common point at $\rho = 0$, of 2400 S cm^2 mol^{-1}, which is somewhat greater than that of 1800–1900 S cm^2 mol^{-1} for the other 1 : 1 electrolytes studied so far. Given the substantial error involved in estimating Λ^{∞} at these temperatures and the required pressures, up to 12 kbar, it would appear that 1 : 1 electrolytes have a common maximum value for conductivity that is independent of temperature above 400°C. Furthermore, they exhibit a very similar density dependence (except for HBr). Observations similar to those above, i.e., a linear dependence of Λ^{∞} on water concentration in g cm^{-3}, have been observed for a range of water–dioxane mixtures at 400°C.[43]

At finite concentrations the molar conductivity passes through a maximum with increasing density, the magnitude of which is a function of temperature and concentration Fig. 2-21. At low densities the salt is appreciably associated in water because of the low dielectric constant of the solvent. An increase in pressure results in an increase in ε and, therefore, in the degree of dissociation, so the conductivity increases. However, in competition with this effect is the increase in viscosity, also a result of the increasing density; this tends to reduce the mobility of an ion and reduce the conductivity. The conductivity maximum represents the point at which the two competing effects have the same density dependence. Any decrease in the degree of ionization, and therefore conductivity, caused by increasing the temperature or concentration, will cause the maximum to diminish and shift to higher densities. Not shown in Fig. 2-21 is the effect of temperature on Λ at constant high densities. The conductivity passes through a maximum with increasing temperature at constant

---

† Estimated values.

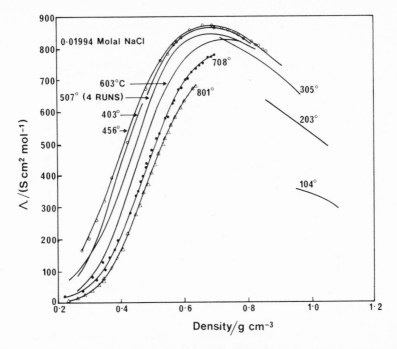

FIG. 2-21. Λ versus density for 0.01994-mol-kg^{-1} NaCl solutions at several temperatures. Reprinted with permission from A. S. Quist and W. L. Marshall, *J. Phys. Chem.* **72**, 684–703 (Fig. 17) (1968). Copyright 1968 American Chemical Society.

high density, Fig. 2-22, and the decrease in Λ at higher temperatures is believed to be the result of a small increase in ion association.

Most unsymmetrical electrolytes exhibit ion association at normal temperatures and pressures and, as might be expected, this has an even more significant effect on the conductivity at high temperatures and low densities. The Λ_{equiv} versus ρ curves for $BaCl_2$, for example, lie below those of a KCl solution of similar concentrations at high densities where ion association becomes less significant.[41] Similar observations have been made on K_2SO_4 solutions to 800°C and 4 kbar.[38a] The Λ^{∞} data for this salt are not very reliable above 300°C, but they exhibit a similar temperature and density dependence to that observed for KCl solutions; Λ^{∞} extrapolates to a value at $\rho = 0$ of approximately 2000 cm^2 ohm^{-1} equiv^{-1}. Λ^{∞} for $BaCl_2$ is also independent of temperature above 400°C and decreases only slightly with increasing density, e.g., $\Lambda^{\infty} \sim 1100$ S cm^2 equiv^{-1} at 0.5 g cm^{-3} and ~ 990 at 0.9 g cm^{-3}.

The conductivity curves for $KHSO_4$ exhibit an unusual density dependence that is clearly the result of progressive ionization.[38c] Above 400°C,

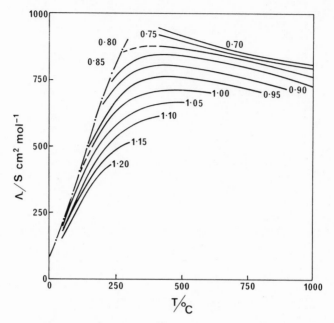

FIG. 2-22. Λ versus temperature of 0.01-mol-cm^{-3} KCl at several densities. Bottom to top: 1.20, 1.15, 1.10, 1.05, 1.00, 0.95, 0.90, 0.85, 0.80, 0.75, and 0.70 g cm^{-3}. Reprinted with permission from K. von Mangold and E. U. Franck, *Ber. Bunsenges Phys. Chem.* **73**, 21–27 (Fig. 3) (1969).

KHSO$_4$ behaves as a 1 : 1 electrolyte because the HSO$_4^-$ ion undergoes very little ionization even at 4 kbar. However, below 400°C and at higher densities the dielectric constant of water is such that this ion will dissociate. The result is an upward turn in the Λ versus density curve at densities greater than 0.8 g cm^{-3}.

2.6. Excess H$^+$ and OH$^-$ Mobility in Aqueous Solutions

The limiting ionic conductivities of H$^+$ (aq) and OH$^-$ ions in aqueous solutions are significantly larger than can be accounted for by any hydrodynamic theory (Fig. 2-23). To explain these observations it has been proposed that H$^+$ aq and OH$^-$ (aq) conductance arises from two mechanisms, a hydrodynamic one and a special proton-transfer mechanism.[65] The hydrodynamic contribution is commonly regarded as being approximately equal in magnitude to that of K$^+$ for the hydrogen ion and Cl$^-$ for the hydroxide ion. If these assumptions are regarded as reasonable first approximations, then the excess conductances, $\lambda_E^{\infty}(H^+)$ $\lambda_E^{\infty}(OH^-)$ attributable to the special mechanism are calculated as follows:

$$\lambda_E^{\infty}(H^+) = \lambda^{\infty}(H^+) - \lambda^{\infty}(K^+) = \Lambda^{\infty}(HCl) - \Lambda^{\infty}(KCl)$$

and

$$\lambda_E^{\infty}(OH^-) = \lambda^{\infty}(OH^-) - \Lambda^{\infty}(Cl^-) = \Lambda^{\infty}(KOH) - \Lambda^{\infty}(KCl)$$

The excess conductivity comprises about 62% of the conductivity of the $OH^-(aq)$ ion and about 79% of the $H^+(aq)$ ion at 25°C, and $\lambda_E^{\infty}(H^+)$ increases with pressure to 5 kbar at a greater rate than $\lambda^{\infty}(H^+)$ itself, although both become less pressure dependent at higher temperatures.[65, 66]

Todheide[67] has summarized all the reliable high-pressure and temperature conductivity data on H^+ and OH^- within the density range of $0.7–1.15$ g cm^{-3} up to 400°C; he identifies the following features:

1. $\lambda^{\infty}(H^+)$ and $\lambda^{\infty}(OH^-)$ show roughly the same behavior in the density and temperature ranges covered.
2. They are both nearly independent of density.
3. At low temperatures they increase with increasing temperature. At about 300°C a maximum is passed, and at higher temperatures $\lambda^{\infty}H^+$ decreases slightly with further rise in temperature. This maximum is an

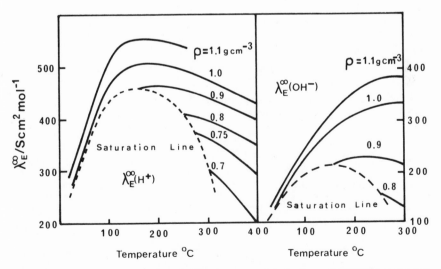

FIG. 2-23. Conductivity of H^+ and OH^- in excess of hydrodynamic limitations; see text for explanation. Reprinted with permission from K. Todheide, in *Water, A Comprehensive Treatise* (F. Franks, ed.), Vol. 1, pp. 463–514 (Fig. 8), Plenum Press, New York (1972).

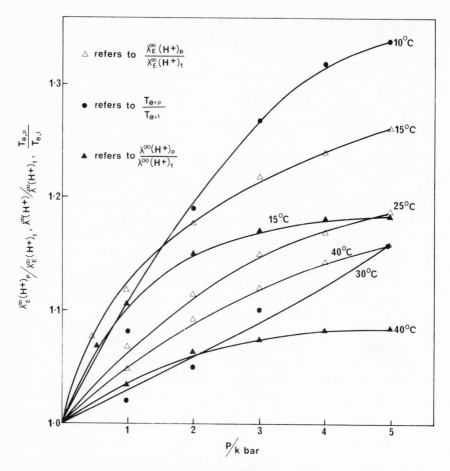

FIG. 2-24. Excess H^+ (aq) molar conductivity $\lambda_E^\infty(H^+)$ and molecular reorientational correlation times τ_θ versus pressure at several temperatures.[67]

unusual feature and must be attributable to $\lambda_E^\infty(H^+)$, which passes through
a distinct maximum with increasing temperature (Fig. 2-23). $\lambda_E^\infty(H^+)$ and
$\lambda_E^\infty(OH^-)$ increase with increasing density over the experimentally accessi-
ble temperature range, being much more density dependent at higher
temperatures.

All the features described above are in qualitative accord with the
model proposed by Eigen and De Maeyer[65] for excess proton mobility.
They propose that the protonic charge is contained within a hydrogen
bonded complex $H_9O_4^+$. The charge can fluctuate rapidly within the
complex, but the mobility of the charge in the direction of the applied field

is determined by the formation and decomposition of hydrogen bonds at the periphery of the complex. These bonds are formed by water molecules formerly in the surrounding water structure, but after liberation from this structure and a suitable reorientation time, they attach to the complex, while other water molecules detach from the other side. In this manner the charged complex moves forward in the direction of the applied field, and the rate-determining step should be the rate at which a water molecule undergoes liberation from adjacent water molecules and subsequent reorientation. In alkaline solutions proton transfer occurs via a defect mechanism within the $H_7O_4^-$ complex.

In Fig. 2-24 are compared the molecular reorientational correlation times τ_θ, from NMR relaxation studies of water, and $\lambda_E^\infty(H^+)$, $\lambda^\infty(H^+)$, as a function of pressure at various temperatures. The pressure dependences of τ_θ and $\lambda_E^\infty(H^+)$ are very similar, suggesting a strong correlation between excess proton mobility and molecular reorientation in water to 5 kbar and 40°C. With increasing temperature at constant density the molecular reorientational relaxation time should decrease, and hence $\lambda_E^\infty(H^+)$ and $\lambda_E^\infty(OH^-)$ should also increase as observed. However, if the molecular reorientation time is more temperature dependent than the rate of proton transfer then at higher temperatures the rate determining step will be the proton-transfer mechanism. But if the transfer mechanism within the $H_9O_4^+$ complex is strongly dependent on the orientation of water molecules, then increasing temperature will disrupt that orientation and cause a decrease in the rate of proton transfer, and in $\lambda_E^\infty(H^+)$, $\lambda_E^\infty(OH^-)$, as observed in Fig. 2-23.

The density dependence of $\lambda_E^\infty(H^+)$ and $\lambda_E^\infty(OH^-)$ up to about 90°C is dependent upon a diminishing reorientational relocation time. However, above this temperature the density dependence becomes much greater, and this may be due to the effect of density on intermolecular distance, the probability of $H_9O_4^-$ complex formation, and the associated proton-transfer rate.

The experimental data for HBr indicate that above 400°C and at the limit of zero density the mobility of H^+ ions is equal to that of other univalent cations,[38f] and this is probably true for OH^- ions as well. However, even at low densities at 400°C, excess proton mobility is evident, and this confirms other experimental evidence as to the existence of hydrogen bonding in high-temperature water.

High concentrations of electrolyte can also affect the excess proton mobility (Fig. 2-25[66]). At compositions greater than 1 $mol\,kg^{-1}$ the reduced mole fraction of water probably inhibits the special proton-transfer mechanism. The errors involved in the approximations leading to the calculation of $\lambda_E^\infty(H^+)$ and $\lambda_E^\infty(OH^-)$ may be responsible for the initial

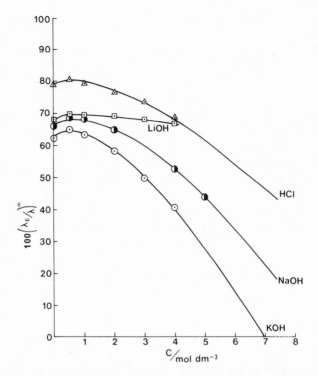

FIG. 2-25. Concentration dependence of the percentage contribution of the proton-transfer mechanism to the conductance of H^+ and OH^- in aqueous HCl, LiOH, NaOH, KOH at 25°C and 1 atm. Reprinted with permission from D. A. Lown and H. R. Thirsk, *Trans. Faraday Soc.* **67**, 132–152 (Fig. 4) (1971).

rise below 1 mol kg^{-1}; however, these errors are likely to be least significant at low concentrations, and the rise appears systematically for the systems studied. If it is a real phenomenon then it may arise from a loosening of the water structure in the vicinity of structure-breaking ions.

Implicit in the above discussion was the fact that $\lambda^\infty(H^+)$ and $\lambda^\infty(OH^-)$ can be determined unambiguously by extrapolating conductance data from finite concentrations to infinite dilution via the Onsager limiting law. Conductance data for HCl, for example, fit the Onsager law to 10^{-3} mol dm^{-3} at 25°C, and either the excess proton conductance is independent of concentration to this level or it follows the same concentration dependence as the hydrodynamic contribution. The hydrodynamic mechanism presumably arises from the downfield drift of a hydrated H_3O^+ (an $H_9O_4^+$) ion, and the dynamic properties of the surrounding solvent molecules that determine the mobility of this entity are probably those that determine the excess proton mobility, i.e., hydrogen bond rupture and formation and molecular reorientation.[68]

Excess H^+ ion conductance has also been observed in methanol,[69] propan-1-ol, and propan-2-ol.[30] Presumably the excess conductance mechanism arises from proton switching across the $R-H_2O^+$ molecules. In propanol solutions $\lambda_E^\infty(H^+)$ increases with pressure at 25°C, reaching a plateau at 1.5 kbar at which it remains, even at 3 kbar.

2.7. The Limiting Ionic Conductivity of Ions in Solution

The theoretical interpretation of the effects of ion size, solvent, temperature, and pressure on the limiting ionic conductivity remains in an underdeveloped state. This may be in some part due to the chemist's preoccupation with the conductance theories of Onsager, Fuoss, Falkenhagen, and others, but it is also to the fact that Λ^∞ is significantly dependent, but to an unknown extent, upon specific ion–solvent interactions as well as the bulk properties of the solvent. Over the last 20 years experimental techniques have been developed to study these effects to the extent where, for water at least, a significant body of data has led to improved understanding (if still at an empirical level) of ion–solvent interactions.

Ideally it would be possible to calculate λ_i^∞ from an appropriate model of the charge-transport process in ionic solution. However, a satisfactory model has not yet been developed for ionic transport in simple systems where specific ion–solvent interactions are absent, e.g., argon or molten KCl. Transition state theory has been tested for aqueous and nonaqueous solutions and provides a useful way of correlating the effects of temperature and pressure on ionic conductance but very little insight into the transport mechanism. In spite of this lack of success, both Brummer[70] and Hills,[71] proponents of transition state theory, have in the past been outspoken in their criticism of Walden's rule (Chapter 1) as a means of interpreting the behavior of λ^∞ as function of r_i, temperature, pressure, and solvent. The fluidity $(1/\eta)$ and the conductance are generally similar functions of temperature and pressure because the mechanism of ionic transport is determined by factors similar to those for viscous flow. Thus the product $\lambda^\infty \eta$ cannot give information on the mechanism of ionic transport but only the difference between this mechanism and that of hydrodynamic flow. It is this fact that makes the study of the Walden product fruitful, because it highlights effects due to specific ion-solvent interactions, the knowledge of which are essential to any successful theory of ionic transport in electrolyte solutions.

2.7.1. The Molecular Hydrodynamic Approach

The limiting ionic conductivity of an ion moving under the influence of an electric field in a viscous continuum is given by Stokes' law as

$$\lambda^\infty \eta = \frac{F|z_i|e}{f \pi r_s} \tag{2.38}$$

where $f = 4$ for true slip and $f = 6$ for true stick. According to this equation, the Walden product, $\lambda^\infty \eta$, is independent of temperature and pressure and should increase linearly with $1/r_s$. That this is not the case is shown in Fig. 2-26 for a range of hydrogen-bonded solvents; similar curves are observed for other solvents.

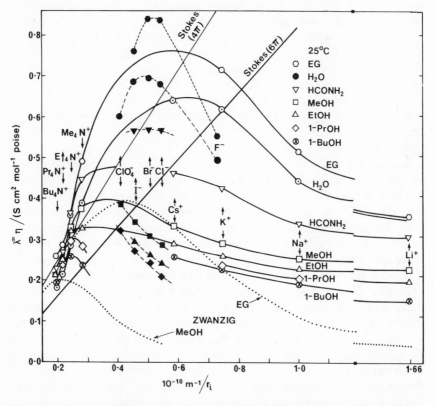

FIG. 2-26. Limiting Walden products for ions in ethylene glycol (EG), water, formamide, methanol, ethanol and propan-1-ol versus reciprocal of ionic radius. Reproduced with permission from R. L. Kay, D. F. Evans, and Sister M. A. Matesich, in *Solute–Solvent Interaction* (J. F. Coetzee and J. F. Ritchie, eds.), Vol. 2, 105–153 (Fig. 10.11), Marcel Dekker, New York (1976).

There are now thought to be at least three phenomena that contribute to the shape of these curves,[72] and they arise from:

(i)   viscous hydrodynamic drag;
(ii) a dielectric friction effect;
(iii) ion–solvent interactions or the "cosphere effect."

The dielectric friction effect, discovered by Born,[73] Fuoss,[74] and subsequently developed by Boyd[75] and Zwanzig,[76] arises from the ability of an ion to orientate solvent dipoles in its cosphere. Because of the mutual attraction of the ionic charge and solvent dipoles and the finite time required for the solvent to relax after the passage of the ion, it experiences a net retardation force. Zwanzig's treatment of this effect is in essence a correction to Stokes' law and is given as

$$\lambda_0 \eta = \frac{F|z_i|e}{f\pi r_i + B/r_i^3} \qquad (2.39)$$

where

$$B = H\phi$$

$$\phi = (z_i e)^2 (\varepsilon_0 - \varepsilon_\infty)\tau / \varepsilon_0 (2\varepsilon_0 + 1)\eta \qquad (2.40)$$

τ is the dielectric relaxation time, ε_0 and ε_∞ the low-frequency and high-frequency dielectric constants, H is equal to $\frac{3}{8}$ and $\frac{3}{4}$ for perfect sticking and slipping boundaries, respectively. With this modification the product $\lambda^\infty \eta$ passes through a maximum with increasing $1/r_i$, as observed experimentally in Fig. 2-26, but with the exception of large r_i values it predicts Walden products that are too small. The discrepancy is greatest for those solvents with a high degree of hydrogen bonding and least for aprotic solvents like acetonitrile and acetone. It also predicts a common curve for both anions and cations, whereas $\lambda^\infty \eta$ differs considerably according to the sign of the charge on an ion. The data for propanol and butanol do not show a single maximum, and the curve for ethanol does not exhibit a smooth change in slope shown by methanol water and ethylene glycol: This phenomenon may arise from a change in conductance mechanism. However, the $\lambda^\infty \eta$ curves for divalent and trivalent ions show approximately the trends predicted by equation (2.39),[77] i.e., the $\lambda^\infty \eta$ maxima increase and shift to smaller $1/r_i$ values with increasing charge, although the predicted values always fall below the observed ones. Thus, while the Zwanzig modification to Stokes' law predicts values of $\lambda^\infty \eta$ that are too low for any given ionic radius, it does predict the general shape of the $\lambda^\infty \eta$ versus $1/r_i$ curves and the relative changes in the position and height of the maximum with increasing ionic charge. A suggestion was made that dielectric saturation of the solvent dipole in the vicinity of the ion would reduce the dielectric frictional force; subsequent calculations, however, showed that this factor increases the discrepancy.[78] Hubbard and Onsager[79] have recently improved the theory of dielectric friction in the electrolyte solutions. They derive an expression for the friction coefficient that is similar in form to Zwanzig's earlier equation but in this case for both the slip and stick boundary condition. Their analysis results in an expression for B in equation (2.39) given by

$$B = H\phi$$

where

$$\phi = \frac{(z_i e)^2 \tau}{\eta} \frac{\varepsilon_0 - \varepsilon_\infty}{\varepsilon_0^2}$$

and $R = 17/280$ for stick and $R = 1/15$ for slip. From this expression, and the appropriate solvent parameters, equation (2.39) predicts that for perfect slip the lower-bound values of the Walden product are 0.19 for Li^+, 0.45 for Na^+, and 0.58 for K^+. The maximum corresponds with the radius of the potassium ion, and the Walden product approaches the Stokes' law line in Fig. 2-26 at $r = 2.5$ Å, 0.4 Å^{-1}. At face value this represents a substantial improvement to the theory of coupled viscous flow and dielectric friction and may represent about the best that can be achieved with continuum models.

Quantitative predictions have been made of the temperature and pressure coefficients of the Walden product.[33] From equation (2.39)

$$\left[\frac{\partial(\lambda^\infty \eta)}{\partial T} \right]_P = F|z_i|e \left(\frac{\partial \phi}{\partial T} \right)_P H / \left[r_i^3 \left(f \pi r_i + H\phi r^{-3} \right)^2 \right]$$

and if the temperature dependence of τ is given by $\tau = 4\pi d^3 \eta / kT$ and $\varepsilon_0 \gg \varepsilon_\infty$, then

$$\left(\frac{\partial \phi}{\partial T} \right)_P \simeq -\phi \left[\frac{1}{T} + \left(\frac{\partial \varepsilon_0}{\partial T} \right)_P / \varepsilon_0 \right]$$

where d is the radius of a water molecule. The calculated values of the temperature coefficient, at two pressures, along with some measured values are shown in Fig. 2-27. A shallow minimum is predicted but at a slightly lower radius than that suggested by the experimental data, and furthermore the measured temperature coefficients are significantly larger than predicted.

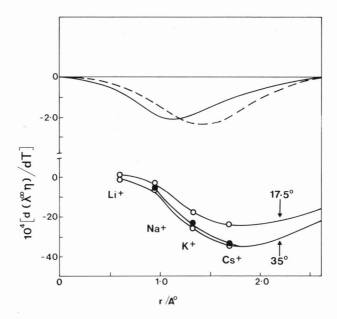

FIG. 2-27. Predicted and measured temperature coefficients of $\lambda^{\infty}\eta$ at 25°C and 1 atm. Solid curve, predicted for perfect stick; dotted curve predicted for perfect slip. Reproduced with permission from E. Inada, *Rev. Phys. Chem. Jap.* **46**, 19–30 (Fig. 8) (1976).

The pressure coefficient of $\lambda^{\infty}\eta$ can be derived from equation (2.39) in a similar manner; it should be positive and exhibit a maximum at $r_{max} = (3H\phi/5f\pi)^{1/4}$. The measured pressure coefficients are negative, except at the highest temperatures and pressures; they do, however, pass through a maximum, Fig. 2-28. The trend toward predicted behavior at higher temperatures and pressures may be a reflection of the fact that the peculiar structural properties of water are considerably diminished at 1000 bar and/or 40°C, and ion–solvent interactions will have less influence on the local viscosity at the surface of an ion. The contribution that ion-solvent interactions make in determining the mobility of an ion will be discussed in some detail below, but there is a possible contribution to deviations from Stokes' law that arises from the arbitrariness of applying the theoretically determined stick and slip coefficients to ions in solution. A way of determining these coefficients experimentally is to measure the diffusion coefficient of uncharged molecules and from the expression

$$D\eta = kT/f\pi r$$

determine the coefficient f for a molecule of radius r. The results obtained for a series of diffusion experiments with the tetraalkyltin ions, that

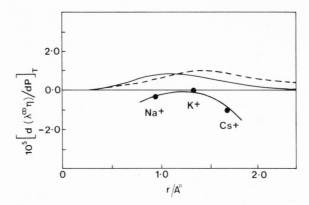

FIG. 2-28. Predicted and measured pressure coefficients of $\lambda^\infty \eta$ at 25°C and 1 atm. Solid curve, predicted for perfect stick; dotted curve predicted for perfect slip. Reproduced with permission from E. Inada, *Rev. Phys. Chem. Jap.* **46**, 19–30 (Fig. 7) (1976).

compare in size and shape with the tetraalkyl ammonium ions, show unexpected deviations from Stokes' law.[80] According to Stokes' law, $D\eta/kT$ should be invariant with temperature. However, this is not the case, and these neutral molecules exhibit even greater departures from the law than the tetraalkylammonium ions in the same solvents. The deviations may arise from interaction between the solvent molecules and the long hydrocarbon chains of the higher alcohols.

Ion–solvent interactions and their effect on the mobility of ions in solutions are generally discussed in terms of the Frank–Wen model of ionic solutions.[81] This model envisages an ion in water as being surrounded by three concentric regions. The innermost region (region A), formed around small or multi-charged ions, contains water that is electrostrictively hydrated, and these molecules tend on average to move with the ion. The second (region B) contains a disrupted water structure and beyond that a region (region C) of normal water. For an ion of high surface charge density (e.g., Li^+) region A might be significantly greater than B, thus an ion surrounded by electrostrictively bound water would have a mobility lower than that predicted from Stokes' law (negative excess mobility) for an unhydrated ion of the same radius. On the other hand, an ion of low surface charge density (e.g., I^-) would have region B greater than region A. The ion would move through a region of disrupted water structure and hence lower viscosity. The mobility of the ion would therefore be in positive excess of that predicted from Stokes' law and the bulk viscosity. In fact for I^- in glycerol (another H-bonded solvent) NMR relaxation studies indicate that region A is nonexistent, i.e., that the region of disrupted structure is directly in contact with the I^- ion.[82] A further effect has been postulated for large ions containing hydrophobic side chains (e.g., tetraalkylammonium ions), that of hydrophobic hydration. It is envisaged that for these ions the center of ionic charge is sufficiently remote from the side chains that it does not alter the normal water structure, nor do the side chains themselves interact with water molecules. Thus water molecules at the surface of the ion are more likely to be orientated into a favorable water structure by neighboring water molecules. This surrounds the ion with a clathratelike structure that tends to reduce its mobility by increasing the local viscosity as well as increasing the size of the moving entity. Detailed discussions of the structure breaking and making effect of ions in solution can be found in references 83 and 77.

The effect of the above phenomena on the mobility of ions in solution is highlighted by studying the change in the limiting ionic Walden product with temperature and pressure. In Fig. 2-29 are displayed the Walden product temperature coefficients for ions in water and D_2O.[84] As discussed above, the temperature coefficients for K^+ and Cs^+ ions are more negative than predicted by equation (2.39), and the following discussion in

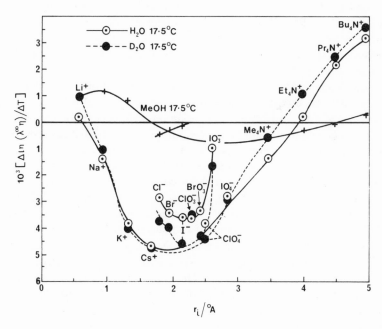

FIG. 2-29. Temperature coefficients of conductance for various ions in H_2O, D_2O, and MeOH at various temperatures. Reprinted with permission from T. L. Broadwater and R. L. Kay, *J. Solution Chem.* **4**, 745–762 (Fig. 8) (1975). Methanol data from Ref. 72.

terms of ion–solvent interaction accounts qualitatively or the large negative values of these ions as well as others in Fig. 2-29.

Within the probable limits of experimental error Li^+ has a zero temperature coefficient while that of the Na^+ is also very small. This is as predicted from equation (2.39); furthermore, since these ions are surrounded by a layer of electrostrictively bound water molecules that are not disrupted by small increases in temperature, the effective hydrodynamic radii would remain constant. There would be no change in the ion–solvent contribution to ionic mobility with increasing temperature. The larger alkali metal ions—the halide, halate, and perhalate ions—all exhibit negative temperature coefficients. This is to be expected of structure-breaking ions since their positive excess mobility would decrease as the structure of bulk water approaches that of cosphere solvent as the temperature is raised. As anticipated, this effect is enhanced when these ions are placed in D_2O, a more structured solvent than H_2O. In water at 17.5 and 35°C, Et_4N^+ has a zero temperature coefficient, but it becomes positive in D_2O; whereas Me_4N^+ has a negative coefficient in H_2O, which becomes zero in D_2O. The observed shifts reflect the changes in the balance of the structure-breaking characteristics of these ions with their hydrophobic structure-making characteristics, the latter being enhanced in the more structured D_2O. Pr_4N^+ and Bu_4N^+, however, exhibit strong structure-making properties, and their temperature coefficients are positive and increase with ionic radius [positive values are not predicted by equation (2.39)]. These observations can be explained by assuming that their negative excess conductances are reduced as increasing temperature reduces the excess water structure around the ions.

However, if this model is correct then in order to explain the effect of pressure on $(\lambda^\infty \eta)$ of tetraalkylammonium ions, Fig. 2-30, it is necessary to assume that increased pressure causes an increase in hydrophobic hydration such that the excess conductance becomes more negative with increasing pressure. As might be expected by comparison with the temperature coefficients of $\lambda\eta$, the apparent increase in hydrophobic hydration increases with increasing size or surface area of the ion since the Walden product decreases with increasing ion size at any given pressure. This trend is also observed for the series $Me_nNH_{4-n}^+(n = 1, 2, 3)$, the series CH_3COO^-, $C_2H_5COO^-$, $C_3H_7COO^-$, and, to a lesser extent, for the series $Et_nNH_{4-n}^+(n = 1, 2, 3)$.[47, 85] The trend seems clearly established; Kay, however,[72] has observed an opposite size dependence, i.e., $(\lambda^\infty \eta)_p/(\lambda^\infty \eta)_1$ falls in the order

$$Pr_4N^+ > Et_4N^+ > Me_4N^+$$

and it is not clear where the reason for the discrepancy lies. However, both

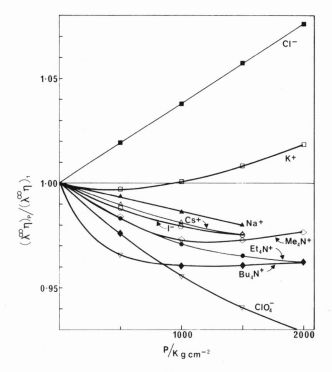

FIG. 2-30. $\lambda^{\infty}\eta$ versus pressure for ions in aqueous solution at 25°C. Data from Refs. 33 and 47.

Kay and the Japanese school at Kyoto University agree that the process of hydrophobic hydration must involve a decreasing volume, i.e., that water in the "hydrophobic sphere" must be more closely packed than in the bulk state. This is a puzzling phenomenon in view of the proposed structure of water in the hydrophobic hydration layer but one that concurs with the estimated negative contribution of this effect to the partial molar volume of the ion at infinite dilution.[86]

The negative pressure coefficients observed for the Na^+, K^+, and Cs^+ ions (Fig. 2-30) can be explained in terms of a reduction in their positive excess mobility that results from a breakdown in water structure at high pressure. Thus the structure of cosphere water, through which the ion is moving, becomes comparable to that of high-pressure bulk water and the ion loses its excess mobility. The sodium ion is noticeably out of place among the alkali ions, and while the observed order is predicted by the Stokes–Zwanzig equation, an alternative explanation proposes that it is due to increased hydration with increasing pressure.[72]

Small hydrated ions (e.g., Li^+, F^-) have large positive pressure coefficients, and Horne has attributed this to a decrease in their hydrodynamic radii as water of hydration is stripped off with increasing pressure.[87] However, it is known that electrostrictive hydration results in a decrease in volume and is therefore favored by an increase in pressure. Kay[72] prefers to explain the excess mobilities of small ions in terms of a pressure induced structure-breaking concept. Thus small ions (e.g., Li^+, F^-) that are electrostrictively hydrated in solution behave as structure breakers in a compressed solvent. The ion would be observed to have a positive excess mobility that increased with pressure. This interpretation is supported by the results of high-pressure NMR spectroscopy on concentrated aqueous solutions.[88, 89] At atmospheric pressure the proton spin–lattice relaxation times T_1 for 4.3 m LiCl, 4.5 m $CaCl_2$, and 4.5 m $LaCl_3$ are less than that of pure water, and this is attributed to the ability of the cations to restrict the mobility of water molecules and increase the relaxation rate. When pressure is applied to these solutions the spin–lattice relaxation time increases to a maximum at 1500–2000 bar. This implies that the solvent molecules in these solutions become more mobile at higher pressures, a kind of structure-breaking effect. Thus small high surface charge density cations appear to have structure-breaking properties with increasing pressure. It is, however, difficult to reconcile this concept with the zero or negative temperature coefficients exhibited by those ions that have positive pressure coefficients.

Negative pressure coefficients have also been attributed to changes in the hydration number with pressure,[27] but these numbers were calculated from Stokes' law in the manner suggested by Robinson and Stokes.[90] This method has now been discredited because it is based on a technique that assumes that the large tetraalkylammonium ions are unhydrated and exhibit no cosphere effects, an assumption contrary to new experimental evidence.[46]

As water is heated, at constant density $\rho = 1$ gcm^{-3}, the ratio of broken-to-unbroken hydrogen bonds drops from $\sim 8:1$ at $10°C$ to $\sim 1:1$ at $400°C$, and to $\sim 0.5:1$ at $\rho = 0.4$ gcm^{-3} at $400°C$. The retention of a significant degree of hydrogen bonding, and presumably some structural features, would lead to the prediction that the excess mobilities shown by ions at ambient temperatures would diminish slowly as the temperature of a solution increased. This prediction is verified by Smolyakov[91] who has shown that the temperature coefficient of the Walden product for the structure-breaking ions K^+, Br^-, Cl^-, and Me_4N^+ remain negative at $200°C$, while those for the electrostrictively hydrated ions Li^+, Ba^{2+}, and Mg^{2+} and the hydrophobic structure-former Et_4N^+ remain approximately constant. In keeping with the previous interpretation of these phenomena at 25 and $35°C$, it is concluded that K^+, Br^-, Cl^-, and

Me_4N^+, for example, retain their structure-breaking characteristics and that Li^+, Ba^{2+}, and Mg^{2+} retain their hydration shells up to at least 200°C. It is perhaps surprising that even the hydrophobic structure-making ion Bu_4N^+ still retains its negative excess mobility, as interpreted from the positive temperature coefficient of $\Lambda^\infty\eta$ at 200°C.

At very high temperatures it would be expected that the abnormal structure of water would be completely broken down, the excess mobilities of ions would tend to zero, and the Walden product would tend to a constant value. Up to 400°C the temperature coefficient of the Walden product for KCl, NaCl, and NaI drops rapidly, Fig. 2-31, after which above $\rho > 0.5$ g cm^{-3} it tends toward a common value independent of temperature and density (a tentative conclusion, but reasonable within the limits of experimental error up to 10%). Below 0.5 g cm^{-3} the product is clearly independent of temperature within the experimental range but decreases with decreasing density. The initial decrease in $\Lambda^\infty\eta$ for KCl is greater than that for NaCl, as might be expected because of its greater structure-breaking ability in low-temperature water; similarly $\Lambda^\infty\eta$ for NaCl and NaI are in keeping with the relative structure-breaking abilities of Cl^- and I^- ions. The data for KCl and NaCl merge above 100°C up to $\rho = 0.7$ g cm^{-3}, beyond which the KCl data becomes less reliable; the

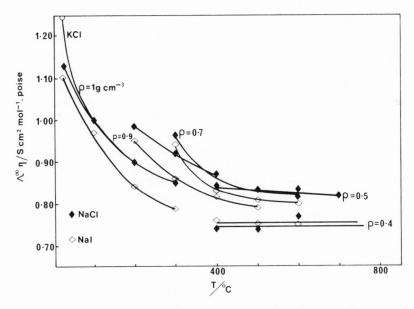

FIG. 2-31. $\Lambda^\infty\eta$ versus temperature for NaCl, KCl, NaI in aqueous solution at various densities. Data from Refs. 37, 38e, and 39.

Walden product for NaI lies below that for NaCl at low temperatures. However, they all exhibit the tendency to extrapolate toward a common value at high temperatures. In terms of Stokes' law this observation could be explained by assuming that in supercritical solutions the thermal energy of solvent molecules is sufficient to overcome most of the solute–solvent interactions and that the mobility of the ions is determined mostly by the viscosity of the solvent and radius of the ion. However, in order to rationalize the measured Walden products at 600°C and $\rho = 0.7$ and 0.4 $g\,cm^{-3}$ with the values calculated from Stokes' law it is necessary to assume that the ions are hydrated so as to reduce their mobility below that expected of a bare ion. Some independent evidence exists to support this conclusion.[86, 92]

The application of the molecular hydrodynamic approach to the interpretation of ionic mobility in other solvents is almost nonexistent because the necessary transport, spectroscopic, and thermodynamic studies have not been conducted. Some data exists for methanol solutions, however, as shown in Fig. 2-26 the Walden products for ions in methanol, predicted from the Stokes–Zwanzig equation, are too low, and they do not account for the different behavior of anions. These deviations can be attributed to ion-solvent interactions, but because of the dearth of spectroscopic and other studies it is difficult to relate these to specific changes in solvent structure in the vicinity of the ion.

The temperature coefficient of $\lambda^{\infty}\eta$ for ions in methanol is displayed in Fig. 2-29. They are noticeably less than those calculated for the same ion in water. If the temperature coefficients arise predominantly from temperature-induced changes in solvent–solute interactions, then such changes must be quite small compared to those that occur in water. This conclusion is complemented by the experimental observation that standard partial molal heat capacities of electrolyte solutions $C_{p,i}^{\infty}$ exhibit less negative values for alkali and halide ions in methanol and less positive values for tetraamylammonium ions in methanol than in water. The large negative $C_{p,i}^{\infty}$ values observed for ions in water can be attributed to structure breaking, the positive values for tetraalkylammonium ions to structure making; however, a completely different interpretation must apply for methanol. In methanol, $C_{p,i}^{\infty}$ increases from Me_4N^+ to Bu_4N^+ and this has been interpreted as meaning that tetraalkylammonium ions are desolvated as temperature is increased, the Bu_4N^+ being initially the most solvated.[93,94] If this were the case, then the Walden product of these ions might show positive temperature coefficients, especially for the largest ones.

For the Br^- ion, $C_{p,i}^{\infty}$ is negative and decreases with increasing temperature. An interpretation of this result is that the number of bonds between solvent molecules in the bulk solvent decreases at a faster rate than the number between solvent and solute as the temperature increases, i.e., bromide ion competes more favorably for solvent molecules as temperature increases, and this process accelerates with increasing temperature. Increased solvation with increasing temperature could reduce the mobility of the ion below that expected and result in a negative Walden product temperature coefficient, as observed.

The heat capacities of transfer $\Delta C_{p,i}^{\infty}(H_2O \rightarrow MeOH)$ for LiCl and LiBr are positive and suggest that quite different Li^+ –solvent interactions occur in methanol and water.[95] In fact, the positive value has been interpreted as being due to the reduced energy of solvation of Li^+ in methanol compared to water because of steric hindrance from the larger methanol molecule. A similar effect has been proposed for Na^+ in methanol. The positive $\lambda^{\infty}\eta$ temperature coefficients for Li^+ and Na^+ in methanol may result from desolvation with increasing temperature.

In contrast to the Walden product temperature coefficients, the Walden product pressure coefficients for most ions in methanol show very large changes with pressure, in most cases considerably greater than those for the same ion in water.[72] The large positive values for the alkali and halide ions have been attributed to pressure-induced structure breaking, but this concept is not well developed. However, if rising temperature can cause an increase for the Br^- –solvent interaction, relative to solvent–solvent interaction, then an increase in pressure may have the opposite effect. In which case the effective ionic radius would decrease with increasing pressure, causing an increase in $\lambda^{\infty}\eta$. This concept, however, is rather naive and cannot account for the behavior of Li^+ and Na^+ ions that, according to the above argument, should exhibit negative Walden product pressure coefficients.

The large Bu_4N^+ ion has an almost zero pressure coefficient indicating that the change in solvation of the ion with pressure is negligible, a conclusion already drawn by Skinner and Fuoss from their studies in methanol at 31°C.[60] By comparison with $Bu_4N \cdot BPh_4$, which had a zero pressure coefficient, they concluded that the pressure coefficients of the Walden products of Bu_4NBr, Me_4NBr, and Na Br were due mostly to that of the Br^- ions. Given this assumption the order of pressure coefficients is similar to that observed by Watson and Kay,[96] i.e., $Bu_4N^+ < Me_4N^+ = Na^+ < Br^-$. A similar order was also observed for the same ions in ethanol, although the Walden products themselves were lower for a given ion at a given temperature and pressure. This trend continues with propan-1-ol and propan-2-ol where the Walden product of Me_4NBr continues to decrease with increasing solvent viscosity and/or molecular mass, Fig. 2-16.

The tetraalkylammonium iodide salts, Me_4NI, Et_4NI, and Pr_4NI, in acetone exhibit large positive Walden product pressure coefficients that are almost independent of the cation,[31] suggesting that the I^- contributes most to the pressure coefficient. The positive excess mobility shown by the I^- ion with increasing pressure may arise from desolvation, the cations being relatively unsolvated and therefore not as susceptible to pressure changes. The temperature coefficients of $\lambda^\infty \eta$ for individual ions in acetone are not known, but those for the Me_4NI and Et_4NI and Pr_4NI suggest that I^- may have a negative coefficient and that for Pr_4N^+ is positive. This conclusion is drawn from the observation that the temperature coefficients become less negative from Me_4NI to Et_4NI, being zero for Pr_4NI. If this conclusion is correct, then these ions in acetone would exhibit similar behavior to the same ions in methanol and water.

The mobility of ions in mixed solvents show some interesting departures from simple Stokes' law predictions. The Walden product of alkali and halide ions in ethanol–water[97] and t-butanol–water mixtures,[98] relative to pure water, $\lambda^\infty \eta / (\lambda^\infty \eta)_{H_2O}$ passes through a maximum at about 10-mol% alcohol. In dioxane–water mixtures the maximum occurs over a range of values and is much less pronounced.[99] Tetraalkylammonium ions in ethanol–water mixtures exhibit a different composition dependence, and after reaching a maximum value at about 10% alcohol the Walden product becomes essentially independent of the organic component.

These observations cannot be explained by dielectric relaxation nor by the usual structure-breaking or structure-making arguments because the height of the maximum decreases in the order $Na^+ > K^+ \simeq Cs^+ > Li^+$ and for anions $ClO_4^- > Cl^- > Br^- > I^-$. The explanation for these observations has been attributed to the phenomenon of "sorting"—the enrichment of one of the mixture components in the ionic cosphere. Thus if water is present in the cosphere to a greater extent than in the bulk mixture, the ion will be moving through a medium of lower viscosity and will exhibit a positive excess mobility. This effect should be greatest for small ions, which except for Li^+ that is probably solvated, is in agreement with experiment. Calculations based on the viscosity of water–ethanol mixtures indicate that the maximum in the Walden product for K^+, Br^- at 10 mol% ethanol can be accounted for by the sorting effect if these cospheres contain 7.2 and 8.8 mol% ethanol, respectively, at 10 and 25°C. The sorting effect should be relatively independent of temperature over a small range, and the observed temperature dependence results therefore from the temperature dependence of the viscosity. Beyond the maximum, the Walden product decreases continuously to the value for pure ethanol.

The positive excess mobilities of the tetraalkylammonium ions, with respect to water, over the whole composition range can be explained by the gradual elimination of the hydrophobic effect as water is removed from the cosphere.

The hydrodynamic molecular approach has proved to be a useful pseudotheoretical framework for coordinating numerous transport, thermodynamic, and spectroscopic studies of electrolyte solutions into a coherent account of the factors that determine ionic mobility in aqueous solutions. Clearly much remains to be explained, especially at high pressure and in nonaqueous solvents.

2.7.2. The Transition State Theory

There have been several applications of the theory of transition states to electrical conductance, but the most preferred is that due to Bockris *et al.*[100] The complete derivation of their equation is given in Chapter 3. The theory is based on the assumption that an ion migrates through a liquid by a series of jumps from one equilibrium position to another and that each jump requires a partial molar Gibbs energy of activation $\Delta^{\ddagger}G_i$ characteristic of the system. Following the procedure of Bockris *et al*, Brummer and Hills[70] derive the equation

$$\lambda_i^{\infty} = \frac{1}{6}\frac{e_i F}{h}(2d_i)^2 \exp\left(\frac{-\Delta^{\ddagger}G_i^{\infty}}{RT}\right) \tag{2.41}$$

where

$$\Lambda_i^{\infty} = \sum_i \lambda_i^{\infty} \tag{2.42}$$

and $2d_i$ is the distance between adjacent equilibrium positions in the liquid, e_i the charge on the ion, the superscript on $\Delta^{\ddagger}G_i^{\infty}$ refers to quantities determined at infinite dilution. By analogy with equilibrium thermodynamics

$$\Delta^{\ddagger}G_i^{\infty} = \Delta^{\ddagger}H_i^{\infty} - T\Delta^{\ddagger}S_i^{\infty} \tag{2.43}$$

and

$$\Delta^{\ddagger}H_i^{\infty} = \left(\Delta^{\ddagger}U_i^{\infty}\right)_p + P\Delta V \tag{2.44}$$

$$\Delta^{\ddagger}H_i^{\infty} = \left(\Delta^{\ddagger}U_i^{\infty}\right)_{v^{\ddagger}} + \int_v^{v^{\ddagger}}\left[-\left(\frac{\partial U}{\partial V}\right)_T + P\right]dV$$

$$\Delta^{\ddagger}H_i^{\infty} = \left(\Delta^{\ddagger}U_i^{\infty}\right)_{v^{\ddagger}} + (\pi_l + P)\Delta^{\ddagger}V \tag{2.45}$$

but provided that π_l, the internal pressure of the system, is equal to $\pi_l^{\ddagger}$, the internal pressure in the activated state, then $(\Delta^{\ddagger}U_i^{\infty})_v = (\Delta^{\ddagger}U_i^{\infty})_{v^{\ddagger}}$. Any difference in these two quantities is usually assumed to be unimportant.[59,62,70,71]

The activation energy at constant pressure $E_{\Lambda,p}$ and at constant volume $E_{\Lambda,v}$, derived from equation (2.41), is

$$E_{\Lambda,p}^\infty = E_p^\infty = RT^2\left(\frac{\partial \ln \lambda_i^\infty}{\partial T}\right)_p = \Delta^\ddagger H_i^\infty + 2\left(\frac{\partial \ln 2d_i}{\partial T}\right)_p RT^2$$

$$E_p^\infty = \Delta^\ddagger H_i^\infty + \tfrac{2}{3}\alpha RT \tag{2.46}$$

where α is the expansivity of the solution

$$E_{\Lambda,v}^\infty = E_v^\infty = RT^2\left(\frac{\partial \ln \lambda_i^\infty}{\partial T}\right)_v = \left(\Delta^\ddagger U_i^\infty\right)_v \tag{2.47}$$

and the activation volume $\Delta^\ddagger V_i$ is given by

$$\left(\frac{\partial \ln \lambda_i^\infty}{\partial p}\right)_T = -\frac{\Delta^\ddagger V_i^\infty}{RT} - 2\left(\frac{\partial \ln 2d_i}{\partial p}\right)_T$$

$$\left(\frac{\partial \ln \lambda_i^\infty}{\partial p}\right)_T = -\frac{\Delta^\ddagger V_i^\infty}{RT} + \frac{\beta}{3} \tag{2.48}$$

where β is the compressibility of the solution. Finally from (2.45)

$$E_p^\infty - E_v^\infty = \left(\frac{\alpha T}{\beta} + p\right)\Delta^\ddagger V_i^\infty + \frac{3}{2}RT^2\alpha = (\pi_I + p)\Delta^\ddagger V_i^\infty + \frac{3}{2}RT^2\alpha \tag{2.49}$$

Thus, providing λ_i^∞ or, as is usually the case, Λ^∞ is known as a $F(T, P)$ for a given solute–solvent system, then $\Delta^\ddagger V_i^\infty$, $\Delta^\ddagger H_i^\infty$, and $\Delta^\ddagger U_i^\infty$ can be calculated. In principle these quantities could be calculated from the model, but so little is known about the factors that determine them that most effort has been directed toward establishing empirical relationships between the activation parameters, the volume and temperature of the system, and the ionic radius of the solute ions.

Except for aqueous solutions, E_v is generally inversely proportional to the volume of the system but independent of temperature to the maximum experimental temperature of less than $100°C$.[31, 62, 63] For large ions it does show a small dependence on ionic radius; however E_v can generally be regarded as being determined mostly by the solvent with a small contribution arising from specific ion–solvent interactions. An early interpretation of E_v was that it represented the energy required by the migrating species to jump into a prepared vacancy, $\Delta^\ddagger U_{i,j}^\infty$, and that $\Delta^\ddagger H_i^\infty \approx E_p$ was the sum of $\Delta^\ddagger U_{i,j}^\infty$ and the energy required to form the vacancy $\Delta H_{i,h}^\infty$. However, Brummer[62] has shown that if vacancies have a partial molar Gibbs energy of formation $\Delta G_{i,h}^{\infty}$ and if the conductance is proportional to the number of vacancies, then

$$E_p^{\infty} \approx \Delta H_{i,h}^{\infty} + \Delta^{\ddagger} H_{i,j}^{\infty}$$

and

$$E_v^{\infty} = \Delta U_{i,h}^{\infty} + \left(\Delta^{\ddagger} U_{i,j}^{\infty} \right)_v$$

and

$$E_p^{\infty} - E_v^{\infty} \approx (\pi_l + P)\left(\Delta V_{i,h}^{\infty} + \Delta^{\ddagger} V_{i,j}^{\infty} \right)$$

where the subscript j refers the activation parameter to the energy or volume change required for a jump into a prepared vacancy. Crude estimates of $\Delta^{\ddagger} V_{i,j}^{\infty}$ show that it is probably a significant fraction of $\Delta^{\ddagger} V_i^{\infty}$, and therefore the simple distinction between E_p^{∞} and E_v^{∞} described above is not justified. As the solution is compressed, however, both $\Delta U_{i,h}^{\infty}$ and $\Delta^{\ddagger} U_{i,j}^{\infty}$ would be expected to increase with the decrease in free volume available for redistribution of the solvent in the process of hole formation, and during the process of translation from one site to another, Table 2-4. In the absence of structural or specific ion–solvent interaction $\Delta^{\ddagger} U_{i,j}^{\infty}$, and hence $\Delta^{\ddagger} U_i^{\infty}$, would be expected to increase with ion size, and this is observed for some tetraamylammonium salts in acetone and nitrobenzene, Table 2-4, but not in N,N-dimethylformamide. $\Delta U_{i,h}^{\infty}$ would be independent of ion size where $r_i < r_s$, the radius of the solvent molecule.

The configurational entropy theory of transport, due to Adam and Gibbs, predicts that E_v and $\Delta^{\ddagger} V_i$ will tend to large values as the system is compressed at constant temperature, see Chapter 3. While E_v^{∞} behaves as predicted in these systems, $\Delta^{\ddagger} V_i^{\infty}$ does not. In fact, $\Delta^{\ddagger} V_i^{\infty}$ in methanol,[59] nitrobenzene,[63] N,N-dimethylformamide,[62] and acetone[31] is proportional to the volume of the solution and inversely proportional to temperature. For $n\mathrm{Bu}_4\mathrm{NP}_i$ in nitrobenzene, Fig. 2-18, $\Delta^{\ddagger} V_i^{\infty}$ is related to a limiting temperature T_L and volume V_L by

$$\Delta^{\ddagger} V_i^{\infty} = \Delta^{\ddagger} V_{i,L}^{\infty} + \frac{k_v(V - V_L)}{T - T_L}$$

where $\Delta^{\ddagger} V_{i,L}^{\infty} = 8.5 \text{ cm}^3 \text{ mol}^{-1}$ is a limiting activation volume. V_L in this equation has the value of 96 $\text{cm}^3 \text{ mol}^{-1}$, which corresponds with the value of 94 $\text{cm}^3 \text{ mol}^{-1}$ at which $1/E_v = 0$; the volume of solid $n\mathrm{Bu}_4\mathrm{NP}_i$ at 1 atm is 91 $\text{cm}^3 \text{ mol}^{-1}$. However, it is difficult to conceive of a process occurring at the close-packed volume of the solid where $E_v^{\infty} = (\Delta^{\ddagger} U_i^{\infty})_v = \infty$ and

TABLE 2-4. E_V for Tetraalkylammonium Salts in Acetone and Nitrobenzene[31, 62, 63][a]

Solvent volume/ $cm^3 mol^{-1}$	$E_V/cal mol^{-1}$		
	Acetone		
	Me_4NI	Et_4NI	nPr_4NI
69.06	300	540	660
71.14	310	430	600
73.63	300	400	540
77.49	470	520	500
	Nitrobenzene		
	Me_4NPi	Et_4NPi	nPr_4NPi nBu_4NPi
99.62	1840	1876	2097 1592
100.91	1444	1430	1480 1450
102.73	1171	1190	1219 1228
105.80	947	917	783 905
	N, N-dimethyl formamide		
	KPi	Me_4NPi	nBu_4NPi
72	1380	1320	1300
74	1130	1030	990
76	1000	880	850
78	930	820	800

[a]Interpolated from the data in Ref. 62.

$\Delta^{\ddagger}V_i^{\infty} = 8.5$ $cm^3 mol^{-1}$ and not infinity. Activation volumes for ionic conductance in solid electrolytes certainly have finite values, but then E_p and E_v are also finite. It is probable that the extrapolation of $\Delta^{\ddagger}V_i^{\infty}$ to a limiting value is not justified, and that at higher pressures and smaller volumes $\Delta^{\ddagger}V_i^{\infty}$ increases with pressure, as has been observed in concentrated solutions.

The trend toward a limiting low value of $\Delta^{\ddagger}V_i^{\infty}$ is also observed for methanol and nitrobenzene solutions but not in acetone; all these observations refer to measurements restricted to 2 kbar. For large ions, $\Delta^{\ddagger}V_i^{\infty}$ is remarkably insensitive to ion size. In N, N-dimethylformamide, at constant volume, there is a trend toward a limiting value with increasing r_i, Fig. 2-17. This limit is generally expected to be the activation volume for solvent viscosity since for large ions Walden's rule is a good approximation.

The effect of temperature on $\Delta^{\ddagger}V_i^{\infty}$ is well illustrated by the data in Fig. 2-18. At lower densities $\Delta^{\ddagger}V_i^{\infty}$ is more temperature dependent, and this seems general for nonaqueous solutions. The separate effects of temperature and volume have been explained as follows.[62]

An increase in temperature at constant volume reduces the bonding between the ion and the solvent in both the initial and transition states. However, if the transition state is assumed to be more rigid than the initial state, then an increase in temperature will affect the initial state most and counter the electrostrictive binding effects of the ion in the solvent. The result would be a decrease in $\Delta^{\ddagger}V_i^{\infty}$. If the volume is increased at constant temperature, however, then this would decrease the volume of both the initial and transition states; but since the latter is most firmly bound, the initial state is affected most and $\Delta^{\ddagger}V_i^{\infty}$ will increase. This argument relies on the assumption that the transition state is more rigid than the initial state, and the basis of this hypothesis is the negative $\Delta^{\ddagger}S_i^{\infty}$ obtained from equation (2.41). If it is assumed that $2d_i = (V/N)^{1/3}$, then by using the experimental value of $\Delta^{\ddagger}H_i^{\infty}$, $\Delta^{\ddagger}S_i^{\infty}$ can be calculated. A negative $\Delta^{\ddagger}S_i^{\infty}$ certainly implies a more rigid transition state, but it is difficult to reconcile this with positive activation volumes, which implies that this state should have a greater volume than the initial state.

The unusual behavior of the ionic conductance of aqueous solutions at lower temperatures and pressures has been discussed. Below ~ 1500 bar and $40°C$ the positive pressure coefficients of conductance must give rise to negative activation volumes; these, of course, become positive at higher temperatures and pressures. Negative activation volumes can be correlated with a collapse of the water structure around the ion in the transition state. As temperature or pressure is increased, the open structure of the initial state will be partly destroyed while that of the transition state would be relatively unchanged and V_{initial} will ultimately become less than the $^{\ddagger}V$.

Except at low densities, E_p for ionic conductance in water is positive. Therefore when $\Delta^{\ddagger}V_i^{\infty}$ is negative, $E_p < E_v$; however, both E_p and E_v are temperature dependent and at low temperatures E_v increases with increasing volume but behaves normally above $50°C$ and at high pressure.[71] For KCl, Larionov has attributed this phenomenon to the breakdown of residual hydrogen bonding between the hydrated K^+ and Cl^- ions and bulk water molecules as the temperature or pressure is increased. This would result in less energy being required for separation of the hydrated ion from its nearest neighbor prior to a jump into the activated state. Assuming that the transition state was relatively unchanged by temperature or pressure, then this would lead to a decrease in E_v with increasing pressure and temperature.[42]

At high temperatures, greater than $350°C$, E_v for NaCl, NaI, NaBr, and $KHSO_4$ tends to zero at ~ 0.70 g cm^{-3}.[38] It may even become negative at low densities ~ 0.40 g cm^{-3}; but under these conditions the errors in assessing Λ^{∞} are quite large and this interpretation may be incorrect. Coincidentally, the activation energy for the viscosity of water E_{η}, at constant volume, becomes zero at ~ 0.7 g cm^{-3}.

Further discussion of attempts to elucidate the features of the transition state are given in Chapter 4 for molten salts and in Chapter 6 for silicate melts.

A recent attempt has been made to calculate λ_i^∞ for a series of ions in aqueous solution by combining the transition state theories for ionic conductance and viscous flow.[101] The measured activation energies for viscous flow and conductivity were used, and the assumption was made that the jump distance in the process of viscous flow is proportional to $V^{1/3}$; the jump distance for ionic translation was assumed to be proportional to the mean distance between centers of ions and the hydration water molecules. The latter can be calculated from X-ray data on aqueous solutions. Not surprisingly, the final equation, which also includes the measured viscosity, is able to reproduce the conductance data to within 3% for KCl and LiCl from 25 to 150°C and to 8 kbar.

The transition state theory as applied to ionic conductivity in liquids provides a convenient set of parameters through which it is possible to compare quantitatively the effect of temperature and pressure on ionic conductance. However, the extent to which it gives a real insight into the transport mechanism of ionic charge is unknown. A detailed critique of the theory is given in Chapter 4.

Ionic Conductivity in Low-Temperature Molten Salts and Concentrated Solutions

3.1. The Transition from Dilute Solutions to Molten Salts

The Debye–Hückel theory of electrolyte solutions, based on the primitive continuum model, becomes invalid for aqueous 1 : 1 electrolyte solutions at about 0.001 mol dm^{-3}.[1] Above this concentration the ionic atmosphere of an ion, i.e., its electrical image in the solution, is closer to the ion than its nearest neighbors—a physical absurdity. However, the modified Debye–Hückel theory, with a suitable value of a fits the activity coefficient data up to about 0.1 mol dm^{-3}.[2] Fuoss and Onsager, using a similar argument to that above, consider that their 1957 conductivity equation will be valid up to about 0.02 mol dm^{-3}.[3] Above this concentration the electrical image of an ion (equal to unit charge) approaches closer than $7a$, and large fluctuations would result in the potential at the central ion on the close approach of a near neighbor. This would invalidate a fundamental postulate of the Debye–Hückel theory, that the potential at the surface of the ion is related, by Poisson's equation, to the average charge density surrounding the ion. However, Fuoss's new theory (Chapter 2) is limited to 0.09 mol dm^{-3} for aqueous 1 : 1 electrolytes and 0.002 mol dm^{-3} for 2 : 1 electrolytes.[4] At these concentrations it becomes impossible to define uniquely an ion pair because three ion interactions become significant. In this theory ions that come within $R > a$ are counted as paired and are excluded from the calculation of long-range effects. This assumption improves the approximation involved in the linearization of the Boltzmann factor in the ion distribution function, but the theory is not developed to account for three ion interactions that make an unknown contribution to the conductance.

For LiCl in a dioxane–water mixture ($\varepsilon = 62.25$), Fuoss calculates that $R = 7.0$ Å at 25°C. Now in a 0.1 mol dm^{-3} solution the average separation of ions is 20 Å, so that on average the Gurney cospheres surrounding the ions would be separated by 13 Å, or approximately four water molecules. At 1 mol dm^{-3} the separation would be equivalent to one water molecule. The continuum model, of ions surrounded by their Gurney cospheres, contained within a dielectric and viscous continuum would be at the limit of validity at 0.1 mol dm^{-3} and not at all applicable above this concentration. For 1 : 1 electrolytes in water the transition from dilute electrolyte behavior to molten salts must occur between ~ 0.1 mol dm^{-3} and approximately 3 mol dm^{-3}. There are numerous examples of this transition.

Goldsack, Franchetto, and Franchetto have recently measured the conductivity of seven 1 : 1 alkali and ammonium halides in water from 0.5 to 10 mol kg^{-1} from 0 to 60°C.[5] Their data illustrate that the molal conductivity, when plotted against concentration, exhibits an approximately linear region from 1 mol kg^{-1} to the highest concentrations studied, Fig. 3-1. Below 1 mol kg^{-1}, the conductivity rises rapidly with decreasing concentration. The transition region would appear to lie between 0.1 and 1 mol kg^{-1}, where below 0.1 mol kg^{-1} the concentration dependence is accounted for in terms of classical electrolyte theory (Chapter 2) and above 1 mol kg^{-1} the concentration dependence of conductivity can be explained in terms of the bulk parameters of the solution rather than those of the solvent.

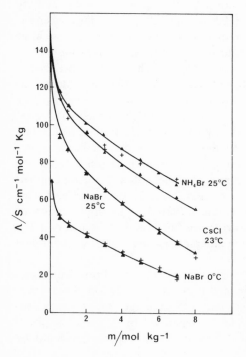

FIG. 3-1. Molal conductivity versus molality for alkali halide salts.[5] ▲ — ▲ experimental results; + calculated from equation derived from equations (3.1), (3.2), (3.3); see Ref. 5.

The pressure dependence of the conductivity also shows a distinct compositional variation from dilute solution to molten salt. Figure 3-2 illustrates how the pressure coefficient $\kappa(c)_p/\kappa(c)_1$ increases linearly with $m^{1/2}$ up to about 0.015 $mol\,kg^{-1}$ (see also Fig. 2-6), then curves over and passes through a maximum in the transition region of ~ 0.1 to ~ 1.5 $mol\,kg^{-1}$. Above 1.5 $mol\,kg^{-1}$ the pressure coefficient decreases with increasing concentration to 10 m, where the mole ratio of salt : water is 1.8 : 1. Aqueous KOH solutions behave in a very similar manner; the pressure coefficient is almost independent of concentration from 0.1 to 1 $mol\,dm^{-3}$ whereafter it decreases linearly with increasing concentration, the rate of decrease becoming greater at higher pressures.[9] Similar trends can be observed from the data for aqueous KCl and NaCl solutions. However, Me_4NCl solutions behave differently, for the pressure coefficient drops sharply to 0.1 $mol\,kg^{-1}$ where measurements cease.[7] The earlier onset of concentrated solution behavior in this case is believed to be due to the large hydration number of Me_4N^+ and therefore the high mole fraction of salt to free water, even in nominally dilute solutions.

In dilute LiCl, KCl, and $CaCl_2$ solutions at high temperatures the conductivity passes through a maximum at constant pressure.[8, 10] The temperature at which the maximum occurs decreases with increasing concentration to 1 $mol\,kg^{-1}$ (Fig. 3-3). T_{max} increases with concentration above 1 $mol\,kg^{-1}$, the limiting value being presumably that of the pure

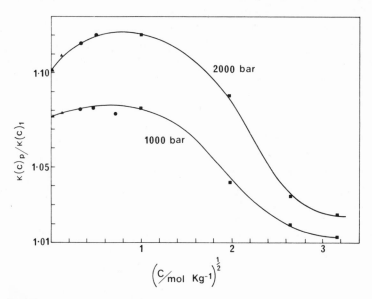

FIG. 3-2. $\kappa(c)_p/\kappa(c)_1$ versus $(m)^{1/2}$ for $LiCl/H_2O$ mixtures at 25°C. Data from Ref. 6–8.

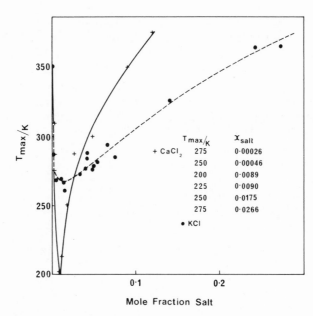

FIG. 3-3. Temperature of conductivity maximum (at constant pressure) T_{max} versus mole fraction of KCl, CaCl$_2$. Data from Ref. 10.

molten salt. A similar trend is observed for aqueous KOH solutions where T_{max} increases for concentrations above 0.1 mol kg^{-1} or $x = 0.0053$.[9] For dilute solutions the conductivity maximum must be caused by the competing effects of increased temperature on the ionic mobility, and the increase in ion association as the relative permittivity of the solvent decreases at higher temperatures. Thus any effect, such as increased concentration or an increase in surface charge density, which would increase the degree of association, would affect the conductance at lower temperatures and cause a decrease in T_{max}.

Above the transition region, 0.1–2 mol kg^{-1}, the conductance maximum is the result of the two competing effects of increased mobility and ion association, but at these concentrations ion association is not controlled by the relative permittivity of the solvent. In pure molten salts it is believed that ion association arises from the formation of gaslike ion pairs in the low-density high-temperature salt.[11] Indeed, Fellows has shown that in the concentrated region, T_{max} follows the same composition dependence as the critical temperature, which illustrates a correlation between the tendency to associate (as shown by T_{max}) and to become gaslike (as shown by T_c).[10]

At the other end of the temperature scale, the ideal glass transition temperatures of a range of Ca(NO$_3$)$_2$/H$_2$O mixtures also illustrates a clear

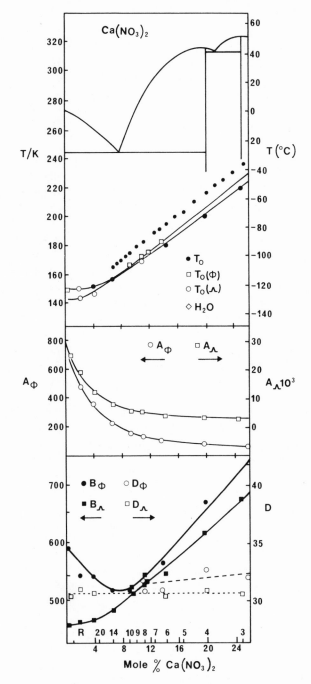

FIG. 3-4. Parameters derived from equation (3.4) versus mole fraction of $Ca(NO_3)_2$ in water. Reprinted with permission from C. A. Angell and R. D. Bressel, *J. Phys. Chem.* **76**, 3244–3253 (Fig. 6) (1972). Copyright 1966 American Chemical Society.

distinction between the properties of solutions above and below 2–3 mol kg^{-1}[12] (Fig. 3-4). The ideal glass transition temperature, T_0, is obtained by curve fitting to the experimental data and reflects the temperature at which the configurational entropy necessary for the transport process becomes equal to zero. For solutions above 4 m, T_0 increases linearly with mole fraction; below 2.3 m it is constant and approximately equal to T_0 for water (Fig. 3-4). The transition region in this case extends from about 2 to 4 mol kg^{-1} and, according to Angell and Bressell, probably has its origin in the incompatibility of the network water structure with the quasilattice structure of a molten salt hydrate. Further evidence of the existence of a boundary or transition region comes from an analysis of T_g for a number of aqueous solutions of mono-, di-, and trivalent metal chlorides and nitrates.[13] T_g is the calorimetrically determined glass transition temperature and falls above T_0, but for $Ca(NO_3)_2$, at least, by a constant amount, Fig. 3-4. The composition dependence of T_g for all the above salts is approximately linear in mole fraction of salt except at low compositions where it becomes independent of composition. T_g becomes independent of composition for trivalent metal chlorides when the mole ratio of water to salt, R, is equal to 24–31, for divalent metal chlorides when $R = 17$–20, and for lithium chloride when $R = 6$–8. These numbers could be regarded as the total hydration numbers for the cation–anion pair and may represent the number of water molecules that the pair can remove from the bulk water structure. In fact, supercooled solutions, whose composition lies in the transition zone, exhibit liquid–liquid phase separation phenomena, the liquid phases presumably being water (or a dilute solution) and a salt-rich hydrated ionic liquid.

3.2. Composition Dependence of Conductance in Concentrated Solutions and Low-Temperature Molten Salts

3.2.1. The Dilute Solution Approach

Some of the equations derived for conductance in dilute solutions can, with suitable modification, be fitted to conductance data of very concentrated solutions. For example, Robinson and Stokes illustrate how a simple form of the conductivity equation (2.2) can be modified to fit conductance data for LiCl to within a few percent up to 9 mol dm^{-3}.[2] This equation, named after Wishaw and Stokes,[14] is for 1 : 1 electrolytes

$$\Lambda = \left(\frac{\eta_0}{\eta} \right) \left(\Lambda_0 - \frac{\beta_0 c^{1/2}}{(1 + \kappa a)} \right) \left(1 + \frac{\alpha_0 c^{1/2}}{1 + \kappa a} \right) \qquad (3.1)$$

where Λ_0, α_0, β_0, κ, and a are defined as before; the last term in parentheses is the relaxation term given by Falkenhagen and others.[15] The viscosity ratio is merely an empirical correction factor, but without this term equation (3.1) fails at quite low concentrations. It fits the data best at high concentrations and works well for nonaqueous electrolytes, e.g., $n(C_5H_4)_4NSCN$ in DMSO[16] and NaI in acetonitrile.[17] The success of this equation in reproducing the experimental data probably lies in the fact that the viscosity factor accounts for most short-range solvent–solute, solute–solute effects, with other deviations being absorbed in the fitting parameter a.

An improved equation for conductance in concentrated solutions has recently been developed,[5] and this is based on the Robinson and Stokes concept of a variable hydration number. It is assumed that the ratio of the solvation number parameter H to the moles of free solvent available to a salt is constant and equal to the ratio at infinite dilution, i.e.,

$$\frac{H}{55.51 - Hm} = \frac{H_0}{55.51} \tag{3.2}$$

where m is the molality in $mol \, kg^{-1}$. Thus the apparent molality m_s, or the number of moles of hydrated solute per mole of free water, is given by

$$m_s = m(1 + 0.018H_0 m)$$

Now if the radius of the solute cation and anion are r_1 and r_2, respectively, and if the number of hydration water molecules are known, then it is possible to calculate the distance parameter a. The result is

$$a_0 = \left[r_1^3 + r_2^3 + 1.25(r_1^2 + r_2^2) + \frac{2.63H_0}{1 + 0.018H_0 m} \right] \tag{3.3}$$

where the third term in the brackets arises from a "dead space" correction due to inefficient packing; the fourth term arises from the contribution of a variable number of water molecules of hydration. The result, when equation (3.1) (without the viscosity ratio) is written in terms of m_s and a_0, is an equation that fits the data for 10 alkali halides and NH_4Cl, NH_4Br in water up to $9 \, mol \, dm^{-3}$. The fit is within several percent at high concentrations, being slightly worse in dilute solutions (Fig. 3-1). In order to obtain this level of agreement between theory and experiment the values of ε and η for pure water were used and ion association was ignored. The change in each of these properties could be accounted for with an

appropriate parameter, but in view of the doubtful validity of the fundamental equation at high concentrations further parameters are unnec-
cessary. Thus the equation should be regarded as semiempirical where the
variable solvation number concept accounts for the variation in conductivity almost as well as the empirical viscosity factor in equation (3.1). The
number H_0 is not necessarily the actual solvation number of the salt in
solution.

Some authors have attempted to use the Walden product as an
indication of the extent of ion pairing in concentrated solutions and
molten salts.[16, 17] In many cases, the product passes through a minimum
with increasing concentration, the minimum moving to higher concentrations with increasing relative permittivity.[18] This observation certainly
indicates that the decrease in the Walden product is due to a dispro-
portionate drop in Λ compared to the associated increase in η. In some
cases the $\lambda^\infty \eta$ value is equal to the $\lambda\eta$ of the molten salt, and this prompted
the hypothesis that the molten salt was therefore fully dissociated and that
for a solution of composition m the ratio $(\Lambda\eta)_m/\Lambda\eta$ is equal to the ion
fraction. However, it has been shown that the Walden product at infinite
dilution contains a significant contribution from specific ion solvent effects
not present in the molten salt. The equality $\Lambda^\infty \eta = \Lambda\eta$ must be mere
coincidence, and furthermore the ratio $(\Lambda\eta)_m/\Lambda\eta$ must also be dependent,
at least in part, on specific solute–solvent effects in the concentrated
solution.

3.2.2. The Molten Salt Approach

An equation that has been successfully applied to concentrated solutions and glass-forming molten salts is the empirical Vogel–Tammann–
Fulcher equation,

$$W(T) = AT^{-1/2} \exp\left[-B/(T - T_0)\right] \qquad (3.4)$$

where $W(T) = \Lambda, D/T, 1/\eta$; A, B are constants for a given transport
function. T_0 is the temperature at which the transport function goes to zero
and has been given theoretical significance by the theories of Cohen and
Turnbull[19] and Adam and Gibbs.[20]

Cohen and Turnbull derived an expression similar to equation (3.4)
based upon the principle that mass transport occurs when, under the
influence of an applied force, a molecule moves into a void of minimum
volume V^* that is created by the redistribution of free volume. Free
volume V_f is that volume over and above that of the close packed liquid

V_0, i.e.,

$$V_f = V - V_0 = \alpha V_m (T - T_0) \tag{3.5}$$

where α is the expansion coefficient, V_m the mean molar volume, and T_0 the temperature at which the free volume becomes zero. The redistribution of free volume is assumed to occur without energy change, and the probability of a void of volume V^* opening up next to a molecule is

$$p(V^*) = \exp\left(-\frac{\gamma V^*}{V_f}\right) \tag{3.6}$$

where γ is a factor to allow for overlap of free volume, $\frac{1}{2} \leqslant \gamma \leqslant 1$. The diffusion coefficient is therefore given by

$$D = gdv \exp\left(-\frac{\gamma V^*}{V_f}\right) \tag{3.7}$$

where $g \approx \frac{1}{6}$, d is the distance through which the molecule travels when it occupies the void volume (effectively a molecular diameter), and v is the velocity of the molecule (the gas velocity). Equation (3.4) is easily obtained from (3.5), (3.6), and (3.7) and the substitution of D for Λ via the Nernst–Einstein equation.[21] This theory is often referred to as the free volume theory (FVT) of Cohen and Turnbull.

The configurational entropy theory developed by Adam and Gibbs considers that molecular translation in a liquid is determined by the probability of cooperative rearrangement in small subsystems. N independent, equivalent, and distinguishable subsystems make up the entire system, and each subsystem contains z molecules, the probability of rearrangement depending on the size of the subsystem. The isothermal isobaric partition function of a subsystem is given by

$$Q(z, PT) = \sum_{E,V} \omega(z, E, V) \exp(-E/kT) \exp(-PV/kT)$$

where ω is the degeneracy of energy level E and volume V of the subsystem. The Gibbs function is given by $G = z\mu = -kT \ln Q$.

Now if n subsystems reside in states that allow a rearrangement, then N/n reside in states that do not, and

$$\frac{n}{N} = \frac{Q'}{Q} = \exp\left(-\frac{G' - G}{kT}\right)$$

where Q' is the partition function corresponding to states in n. The average cooperative transition probability $\overline{W}(T)$ is proportional to n/N the fraction of states that can undergo a transition, therefore,

$$\overline{W}(T) = \overline{A} \exp(-z^*\Delta\mu/kT) \tag{3.8}$$

where z^* is the smallest size subsystem that permits a transition at all.

Now the configurational entropy, S_c, of a subsystem is given by

$$S_c = -(\partial G_c/\partial T)_{p,z}$$

where

$$G_c = -kT \ln Q_c$$

and Q_c is the configurational partition function corresponding to the potential energy part E_{pot} of the Hamiltonian function of the subsystem. The smallest subsystem z^* that can undergo a rearrangement will be characterized by a critical value of the configurational entropy s_c^*; and since the subsystems are independent, the total configurational entropy is

$$S_c = ns_c^*$$

Then for a mole of molecules,

$$N = N_A/z^*$$

and, therefore,

$$z^* = N_A s_c^*/S_c$$

Substitution into (3.8) gives

$$\overline{W}(T) = \overline{A} \exp\left(-\frac{\Delta\mu s_c^* N_A}{kTS_c}\right) \tag{3.9}$$

Now the configurational entropy is given by

$$S_c = \int_{T_0}^{T} \Delta C_p \, d\ln T \tag{3.10}$$

where ΔC_p is the difference in heat capacity between the extrapolated heat capacity of the glass and that of the liquid at temperature T. If ΔC_p is assumed to be inversely dependent on T, then $\Delta C_p = C/T$ where C is a constant.[12] When this equation is substituted into (3.10) and integrated from T_0 to T, the result can be substituted into (3.9) to obtain

$$\overline{W}(T) = \overline{A}\exp\left(-\frac{\Delta\mu s_c^* N_A T_0}{kC(T - T_0)}\right) \tag{3.11}$$

Now the transition probability is proportional to the conductance and since $\overline{A}$ has been shown to have a $T^{-1/2}$ dependence then

$$\Lambda = AT^{-1/2}\exp[-B/(T - T_0)] \tag{3.12}$$

where

$$B = \frac{N_A s_c^* \Delta\mu T_0}{kC} = LT_0 \tag{3.13}$$

Angell and Bressell[12] have shown that (3.12) fits the data for $Ca(NO_3)_2/H_2O$ solutions from 0.1 to 24.8 mol% $Ca(NO_3)_2$ for both conductance and viscosity. B was shown to have identical composition dependence to T_0 thus validating the form of equations (3.12) and (3.13) and, for $Ca(NO_3)_2/H_2O$ at least, the assumption that $\Delta C_p = C/T$.

Equation (3.12) can be written as

$$\Lambda = AT^{-1/2}\exp(LT_0/(T - T_0)) \tag{3.14}$$

and since for $Ca(NO_3)_2/H_2O$ T_0 is linearly dependent on composition, then the isothermal composition dependence of Λ is dependent only on changes in T_0. As T_0 is linear in x, the mole fraction of $Ca(NO_3)_2$, then

$$T_0(x) = T - Y(x_0 - x) \tag{3.15}$$

where x_0 is the composition at which T_0 equals the temperature under consideration, and Y is a constant of proportionality. Equations (3.14) and (3.15) give

$$\Lambda = A'T^{-1/2}\exp[-B'/(x_0 - x)] \tag{3.16}$$

where A', B', and x_0 are constants at any temperature. This equation fits the data for $Ca(NO_3)_2/H_2O$ mixtures over three decades of Λ; however, the fit is not good in the transition region between the molten salt and the dilute solution nor in the dilute solution itself. This undoubtedly arises from the fact that T_0 is not linearly dependent on composition in this region and that A' becomes composition dependent.

While it appears that equation (3.16) has not been tested against the data for other two-component systems, it seems certain that the concentration dependence of other two-component electrolyte solutions will fit (3.16), at least in the range where $T_0 \propto$ composition. A plot of $\ln \Lambda$ versus log (normality), or the logarithm of charge concentration, for aqueous solutions of $LiNO_3$, $Mg(NO_3)_2$, $Ca(NO_3)_2$, $Al(NO_3)_3$, NaOH, and $ThCl_4$ illustrates that the equivalent conductance decreases slowly with log N up to about 3 N and thereafter drops rapidly, tending to zero at about 15–20 N, Fig. 3-5, indicating that conductivity data for all these electrolytes will fit equation (3.16) but with different values of x_0.[22] Furthermore, it seems to be a fairly general rule that T_g is linearly dependent on electrolyte mole fraction in both aqueous and nonaqueous solutions[13] so that T_0 can be expected to behave in a similar manner for most concentrated electrolyte solutions. Mixed salt systems, e.g., $Cd(NO_3)_2 4H_2O / Ca(NO_3)_2 4H_2O$,[23]

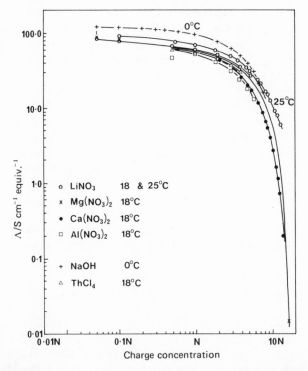

FIG. 3-5. Λ versus charge concentration N for aqueous solutions. Reprinted with permission from C. A. Angell, *J. Phys. Chem.* **70**, 3988–3997 (Fig. 3) (1966). Copyright 1966 American Chemical Society.

$CoCl_2/Ca(NO_3)_2 4H_2O$,[24] appear to exhibit a slightly more complex dependence on composition than equation (3.16) would predict. For the first mixture the best-fit A parameter is observed to be a function of composition; however, small errors, 3%, in the best-fit value of T_0 can introduce variation in A of up to 40% that would easily account for the variation in A with composition. It is not clear whether the observed variation in A is real or merely a result of curve-fitting errors that arise from insufficient data.

For two-component systems at least, the above evidence indicates that the $Ca(NO_3)_2/H_2O$ system is not unique and that the isothermal composition dependence of the conductivity of most concentrated electrolyte solutions will fit on an equation of the form of (3.16). However, for equation (3.16) to be anything more than an empirical expression the constants A', B', and T_0 must be accounted for on the basis of a suitable theory. A discussion of these parameters is given below.

3.3. Temperature and Pressure Dependence of Conductance in Concentrated Solutions and Low-Temperature Molten Salts

Above the transition zone, referred to above, concentrated electrolyte solutions are indistinguishable from ionic melts, at least as far as their transport properties are concerned; for at temperatures not too far above the ideal glass transition temperature, $T < 2T_0$, the conductivity and viscosity of both types of liquid exhibit a similar dependence on pressure and temperature. The conductance falls off with temperature at a rate that is increasingly faster than a simple Arrhenius dependence would predict, Fig. 3-6. This plot shows that the activation energy $E_{\kappa,p}$ [defined by equation (4.6)] increases with decreasing temperature. Angell has shown, however, that this temperature dependence is accounted for by the Adam and Gibbs theory; equations (3.9) and (3.10) predict that E_κ and E_Λ will be a function of T/T_0. Figure 3-7 illustrates that such a relationship does exist for both ionic melts and concentrated electrolyte solutions and, furthermore, that the constant B in equation (3.12) is approximately the same for all the systems in the figure. This figure also quantifies the "low-temperature region" of molten salt transport as being that temperature range where the activation energy becomes significantly temperature dependent and can be arbitrarily established as $T/T_0 \leqslant 2$. Under the condition that $T < 2T_0$, equation (3.12) can be derived from (3.9) and (3.10), but in this case

$$B = \frac{N_A \Delta \mu s_c^*}{k \, \Delta C_p} \tag{3.17}$$

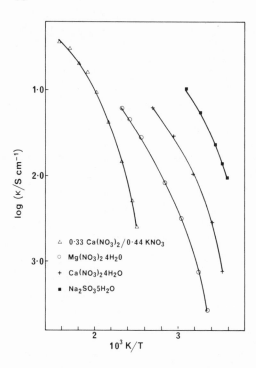

FIG. 3-6. Arrhenius plot of $\log \kappa$ versus $1/T$ for some molten salts.[21, 22, 25]

Now Adam and Gibbs[20] argue that for the case of polymer systems, s_c^* should be nearly the same for all glass-forming liquids, that $\Delta\mu$ is essentially the height of the energy barrier per monomer unit, and that ΔC_p will be proportional to the differences between the minima of the potential energy curve. Evidently, for systems where the intermolecular forces are strong both $\Delta\mu$ and ΔC_p will be large, and conversely for systems with weak intermolecular forces. The ratio of these two quantities would be approximately constant. Applied to molten salts, this argument could account for the apparent constancy of B, as shown for a wide range of similar systems in Fig. 3-7. However, over a wide range B is a composition dependent, Fig. 3-4, and evidently the addition of significant quantities of a nonelectrolyte produces substantial changes in $\Delta\mu$, s_c^*, and/or ΔC_p. It has been suggested that $\Delta\mu$ will be a function of ion–ion interaction energies and effective coordination number,[26] and the reduction in B as water is added to the $Ca(NO_3)_2/H_2O$ system may arise from a decrease in the height of the potential energy barrier without a compensating decrease in ΔC_p. It is also possible that small but systematic differences of the B parameter, between the different systems of Fig. 3-7, are hidden by the errors that arise in the curve-fitting procedure. Small errors in T_0 can cause

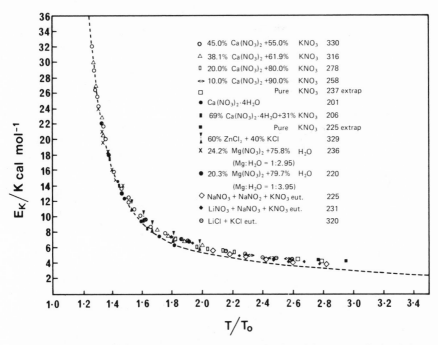

FIG. 3-7. E_κ versus T/T_0 for some molten salts. The dotted line was calculated from equation (3.12), with $B = 680$ K and $AT^{1/2} = $ constant. Reproduced with permission from C. A. Angell, *J. Phys. Chem.* **70**, 2793–2803 (Fig. 2)(1966). Copyright 1966 American Chemical Society.

significantly larger errors in B. There does, however, seem to be a real difference in some of the B parameters calculated from the data for systems corresponding to molten salt hydrates (Table 3-1). It is possible that the larger value of B for the $Ca(NO_3)_2$ system, compared to the $CaCl_2$ system, may arise from a larger value of $\Delta\mu$ and s_c^*. Results from Raman spectroscopy[28] and ^{1}H NMR[29] experiments suggest that inner- and outer-sphere complexes formed between a hydrated cation and a nitrate anion are much more stable than those formed with a Cl^- anion. Lueng and Safford[30] neutron-scattering experiments also indicate that inner- and outer-sphere complexes exist as kinetic entitites and contribute to the diffusion coefficient. Thus the energy barriers to ionic and molecular migration in NO_3^- containing hydrate melts may be greater than those in Cl^- melts. This may result in an increase in $\Delta\mu$ when Cl^- is replaced by NO_3^-, and an increase in s_c^* since the size of the cooperatively rearranging subgroup would also increase. Similar experiments on Mg^{2+} and Ca^{2+} containing hydrate melts show that Mg^{2+} is more strongly bound to its

TABLE 3-1. Values of A, B, and T_0 for Several Hydrate Melts[a]

	P/bar	T_0/K	$A/\text{S mol}^{-1}\text{dm}^2$	$10^{-2}B/\text{K}$	$b/\text{K kbar}^{-1}$
$CaCl_2 \cdot 5.99H_2O$	1	198 ± 2	60.4 ± 3	4.16 ± 0.1	
	2000	205 ± 2	57.4 ± 3	4.09 ± 0.12	
	4000	211 ± 2	59.1 ± 3	4.24 ± 0.14	
	6000	217 ± 2	63.4 ± 3	$4.4 \ \pm 0.2$	
Mean value			60.1 ± 3	4.16 ± 0.15	3.0
$Ca(NO_3)_2 \cdot 3.63H_2O$	1	199 ± 3	2.19 ± 0.09	7.21 ± 0.32	
	2000	206 ± 2	2.26 ± 0.12	7.47 ± 0.24	
	4000	216 ± 3	2.24 ± 0.16	7.44 ± 0.40	
	6000	226 ± 2	2.21 ± 0.13	7.33 ± 0.29	
Mean value				7.36	4.5
$Ca(NO_3)_2 \cdot 4.8H_2O$	1	169 ± 2	2.43 ± 0.07	7.07 ± 0.17	
	2000	181 ± 2	2.34 ± 0.06	6.84 ± 0.18	
	4000	191 ± 2	2.35 ± 0.13	6.85 ± 0.39	
	6000	198 ± 7	2.49 ± 0.28	7.19 ± 0.74	
Mean value				6.99	4.5
$Ca(NO_3) \cdot 6H_2O$	1	180 ± 2		5.40 ± 0.10	4.5
$Mg(NO_3) \cdot 6H_2O$	1	193 ± 5		6.70 ± 0.80	
$CaCl_2 \cdot 5.99H_2O$	1	198 ± 2		4.16 ± 0.15	3.0

[a]Data from Refs. 27 and 36.

water of hydration than is a Ca^{2+} ion and is effectively locked into its cage of hydration water molecules, which may give rise to greater values of $\Delta\mu$ and/or s_c^*. Much more work is required before any definite conclusions can be drawn about the effects of liquid structure on the B parameter.

Probably the most significant parameter to arise from the derivation of equation (3.12) is the ideal glass transition temperature, T_0. This parameter is given thermodynamic significance by the free volume and configurational entropy theories, the latter being now regarded as the most plausible for the reasons given below.

On the basis of "Kauzmann's paradox" it is possible to estimate T_0 from calorimetric measurements, as the temperature at which the entropy of the internally equilibrated liquid equals that of the crystalline phase.[31,32] Kauzmann proposed that since the heat capacity of the supercooled liquid is greater than that of the crystalline phase, it will lose entropy more rapidly than the solid as the temperature is lowered. At some temperature T_0, below the normal melting point, the liquid will have lost entropy equivalent to the entropy of fusion, and at this temperature the heat capacity of the liquid will fall to that of the crystalline solid. This would preclude the formation of an amorphous glassy phase of lower entropy than the crystalline solid, an unlikely occurrence. Thus by measuring the heat capacity of the supercooled liquid and the crystalline and glassy solids it is possible to calculate the temperature at which the entropy S(equilibrated liquid) = S(glassy or crystalline solid). This requires a small extrapolation because the liquid heat capacity usually falls to that of the solid phase at a temperature T_g, some 20°C above the estimated value of T_0. T_g is the experimental glass transition temperature and is the temperature at which the viscous liquid can no longer maintain internal equilibrium as it is cooled at the finite rate determined by the experimental conditions. Figure 3-8 displays the results of such an experiment on $Ca(NO_3)_2 4H_2O$.[33] The predicted $T_0 = 200$ K is very close to that determined from conductance studies (201 K) and from viscosity measurements of 205 K, for $Cd(NO_3)_2 4H_2O$, $T_0 = 194$ K from conductance measurements and 198 K from calorimetric studies. The concurrence of the calorimetrically determined and transport-determined T_0 values strongly support the concept that transport-related structural relaxation times tend to infinity as $S_c = \int_{T_0}^{T}(\Delta C_p / T)\,dT$ tends to zero.

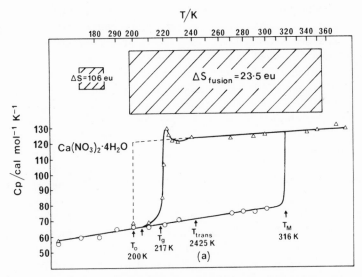

FIG. 3-8. Heat capacity of crystal, glass, and liquid $Ca(NO_3)_2 4H_2O$ versus $\log T$. Also shown in the figure are areas equivalent to the entropies of fusion and solid-state transition of the respective compounds and the temperature T_0 (cal) that satisfies the condition $\Delta S_f + \Delta S_{tr} = \int_{T_0}^{T}[C_p(\text{liq}) - C_p(\text{cryst})]\,d\log T$. Reprinted with permission from C. A. Angell and J. C. Tucker, *J. Phys. Chem.* **78**, 278–281 (Fig. 1) (1974). Copyright 1974 American Chemical Society.

In electrolytic liquids T_0 is dependent on the charge concentration or cohesive energy within the liquid. This is demonstrated for the $Ca(NO_3)_2/H_2O$ system in Fig. 3-4, where T_0 increases linearly with mole fraction above the transition region. The same dependence on composition has been observed for the experimental glass transition temperatures T_g of a number of mono-, di-, and trivalent chlorides and nitrates.[13] T_g in these cases foreshadows T_0 by a constant amount, and the observed composition dependence of T_0 for $Ca(NO_3)_2/H_2O$ systems would be similar to that found in other aqueous systems. For some mixed salt systems with a common anion, Angell[21] has observed that T_0 is linearly dependent on the "cationic strength" defined as $\Sigma_i x_i z_i / r_i$, where x_i is the mole fraction, z_i the charge, and r_i the radius of ion. This approximate relationship extends from salts of weak field cations, e.g. $[Ca(H_2O)_4](NO_3)_2$, to anhydrous ionic solutions. However, it is clear that the cohesive energy of the liquid is not the only factor because in mixed $Cd(NO_3)_2 4H_2O/Ca(NO_3)_2 4H_2O$[23] systems T_0 decreases with increasing Cd^{2+} content; both cations have comparable radii but Cd^{2+} is significantly heavier. Similar results are obtained for T_g, when K^+ is replaced by Tl^+ in $KNO_3/CaNO_3$ mixtures.[34] In this case at about 35-mol% Tl, T_g is 30°C below the equivalent K^+-containing melt.

Very few studies have been made on the anion dependence of T_0 since almost all the low-temperature conductance measurements have been on nitrate melts because of their supercooling ability. Angell and Sare,[13] however, have observed that in concentrated salt–water mixtures, salts with spherical singly charged anions have low T_g's, while asymmetric anions have higher T_g's in approximately inverse proportion to the number of axes about which they can rotate without requiring rearrangements in their nearest neighbor environment. There are exceptions to this observation, e.g., $Ca(BrO_3)_2$, but it is in general accord with the Gibbs and Dimarzio entropy theory of the glass transition for chain polymer liquids and their solutions. The anion dependence of T_g also correlates very well with the pk_a of the acid conjugate to the anion.[13] The pk_a can be regarded as a measure of strength of an anion–water molecule bond and indicates the tendency, in very concentrated mixtures, of the anions to bind protons of water molecules already associated with the cation hydration sheath. This would have the effect of reducing the total configurational entropy and hence raise T_g. Proton NMR studies of concentrated salt–water mixtures support this conclusion by showing that the temperature dependence of the downfield shift, $d\delta/dT$, of the proton resonance increases in an approximately linear manner with pk_a and T_g.[30] The most plausible explanation of $d\delta/dT$ is that it reflects the rate of weakening or breaking of hydrogen bonds between anions and water molecules on the hydrated cation and should therefore correlate with pk_a.

The ideas discussed above are supplemented by the "bond lattice" model of liquid structure.[35] This theory leads to an expression for the degree of configurational excitation of a liquid in terms of the fraction N_x of "bonds" that are broken at a temperature T,

$$N_x = \left[1 + \exp\left(\frac{\Delta H - T \Delta S}{RT} \right) \right]^{-1} \quad (3.18)$$

ΔH is the average value of the energy per mole required to produce a "broken bond" in the liquid (an element of configurational excitation), ΔS is the average molecular change in the liquid lattice entropy accompanying the configurational excitation. A "bond" in this sense can be a potential energy barrier to translation or rotation and not necessarily a covalent or ionic bond. Now if mass transport occurs through cooperative rearrangements of molecules due to an increase in the number of broken bonds at a site within the liquid, then

$$W(T) = 1/z \exp(-\zeta/N_x) \quad (3.19)$$

and

$$W(T) = AT^{-1/2} \exp\left\{ -\left[1 + \exp\left(\frac{\Delta H}{RT} - \frac{\Delta S}{R} \right) \right] \right\} \quad (3.20)$$

where $W(T) \propto \Lambda$ or $1/\eta$, and the constant ζ has been absorbed into ΔS and the constant z into A. This equation can reproduce the experimental data for $Ca(NO_3)_2/H_2O$ with the same precision as the VTF (Vogel–Tammann–Fulcher) equation (3.12).[12] It does not, however, require a sharp transition at a finite temperature T_0 to a configuration ground state where $N_x = 0$. The values for ΔH and ΔS obtained by fitting the conductance data for $Ca(NO_3)_2 4H_2O$ to equation (3.20) are $\Delta H = 2450$ cal mol^{-1} and $\Delta S = 5.08$ cal mol^{-1} K^{-1}. These parameters, when substituted into equation (3.18) and

$$C_{p(conf)} = (\partial H_{conf}/\partial T)_p = R(\Delta H/RT^2)N_x(1 - N_x)$$

give a configurational heat capacity of 5.5 cal mol^{-1} K^{-1} of broken bonds. This is to be compared with the measured value of $\Delta C_p = 48$ cal mol^{-1} K^{-1} of $Ca(NO_3)_2 4H_2O$. Since ΔC_p is believed to arise mostly from the configurational entropy difference between the internally equilibrated liquid and the glass, then there will be $9N_A$ "bonds" per mole of $Ca(NO_3)_2 4H_2O$. These "bonds" may arise from interactions between the protons on the hydrated water molecules ($8N_A$ per mole) and the nitrate anion. Over a range of composition ΔH increases linearly with mol% $Ca(NO_3)_2$, which is consistent with increasing average hydrogen bond strength and increasing inhibition of NO_3^- reorientation.

To test the VTF equation against high-pressure conductance data, it is necessary to perform the experiments over a wide temperature range $\sim 100°C$ and at temperatures not too far above T_0. Very few high-pressure studies have been carried out under these conditions, but the data for several such studies confirm the predictions based on equation (3.12) as derived from the configurational entropy theory. For example, the activation energy, measured at constant volume, $E_{\Lambda,V}$, is predicted by this theory to tend to very large values as $T \to T_0$, as for $E_{\Lambda,p}$. This is in agreement with conductance data for $Ca(NO_3)_2/H_2O$,[36, 37] $CaCl_2/H_2O$,[27] and $KNO_3/Ca(NO_3)_2$ mixtures[38] and for alkyltetrammonium tetrafluoborate salts.[39] The free volume theory, however, predicts that $E_v = -\frac{1}{2}RT$ because the free volume, and hence the exponential term in equation (3.7), remains constant with increasing temperature.

By fitting the data for $KNO_3/Ca(NO_3)_2$ mixtures to the VTF equation Angell et al.[38] have shown that T_0 is linearly dependent on pressure, i.e., $T_0 = T_0' + bP$, where $T_0' = T_0(1\ atm)$. Equation (3.12) now gives

$$\Lambda = AT^{-1/2} \exp\left(- \frac{B}{T - (T_0' + bP)} \right) \qquad (3.21)$$

which, when differentiated with respect to pressure gives

$$\Delta V_\Lambda = -RT\left(\frac{\partial \ln \Lambda}{\partial P} \right)_T = \frac{RTBb}{\left[T - (T_0' + bP) \right]^2} \qquad (3.22)$$

However, since for salt–water mixtures T_0 is also linearly dependent on mole fraction [see equation (3.15)], then

$$\Delta V_\Lambda = \frac{RTBb}{\left[T - (T_0' + bP + Yx) \right]^2} \qquad (3.23)$$

where x is mole fraction of salt, Y is a constant. Equation (3.23) predicts that at constant x and T, ΔV_Λ will tend to large values with increasing pressure; but that at a given P and T, ΔV_Λ will increase with increasing mole fraction of salt. But the most noticeable trend is that predicted at constant x and P where $T \to T_0$. Figure 3-9 for $Ca(NO_3)_2/H_2O$ mixtures shows that all these trends are obeyed. Similar results are observed for $MgCl_2 6H_2O$[37] and $CaCl_2 6H_2O$[36] solutions.

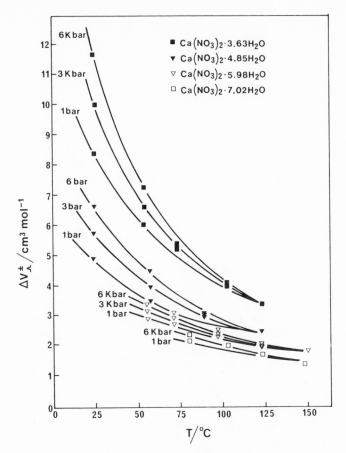

FIG. 3-9. Activation volume ΔV_Λ versus temperature for several $Ca(NO_3)_2/H_2O$ mixtures at 1 bar, 3 kbar and, 6 kbar. Reprinted with permission from L. Pickston, S. I. Smedley, and G. Woodall, *J. Phys. Chem.* **81**, 581–586 (Fig. 3) (1977). Copyright 1977 American Chemical Society.

The data for $Ca(NO_3)_2/H_2O$ mixtures and $CaCl_2 6H_2O$ fit the VTF equation with constant values of B and b (Table 3-1). By comparison with the experimental values of dT_g/dP, the b values determined for ionic liquids follows the expected trend, i.e., they are generally much smaller than those observed for polymeric and network liquids.[42] On the basis that T_0 is the temperature at which the internally equilibrated liquid undergoes a second-order transition ($\Delta S = 0$) to a glass, then it can be shown that

$$dT_0/dP = T_0 V_0 \Delta\alpha/\Delta C_p \qquad (3.24)$$

where V_0 is the molar volume at T_0.[40] A similar expression can be derived for dT_g/dP, but in this case the experimental glass transition must occur at a constant value of the excess entropy, defined by equation (3.10).[41] Equation (3.24) correctly predicts dT_0/dP from the measured expansion coefficients and heat capacities of a number of liquids.[42] $ZnCl_2$ is a notable exception, and the failure of (3.24) in this case has been attributed to the small value of ΔC_p associated with the network liquid.[42]

The apparent pressure independence of B, Table 3-1, was not expected since, from equation (3.13), B should increase with T_0 and hence with pressure. However, the constant $C = \Delta C_p T$ in (3.13) is likely to be pressure dependent; thus $\Delta C_p = C(P)/T$, in which case B could be independent of pressure.[36]

3.4. Electrical Relaxation in Glass-Forming Molten Salts

Conductivity and shear relaxation studies have been made on a number of glass-forming ionic liquids and have provided a useful insight into low-temperature relaxation mechanisms. The electric modulus formalism seems to be the most preferred method of analysis and is based on the principle that if the electrodes are instantaneously charged at time zero and maintained at constant charge, then the electric field E that arises will decay to zero due to the migration of ions.[43] The decay of E is given by

$$E(t) = E(0)\phi(t)$$

where $\phi(t)$ is a decay function of the form

$$\phi(t) = \int_0^\infty g(\tau_\kappa) \exp(-t/\tau_\kappa)\, dt$$

where τ_κ is the electric field or conductivity relaxation time and $g(\tau_\kappa)$ is a normalized density function of relaxation times. In the frequency domain this process may be described by the electric modulus

$$M = \frac{1}{\varepsilon^*} = M' + iM'' = M_s\left\{1 - \int_0^\infty dt \exp\left[(-i\omega t)\left(\frac{-d\phi(t)}{dt}\right)\right]\right\}$$

$$(3.25)$$

where ε^* is the complex permittivity, ω the angular frequency, and

$$M = M' + iM'' = \varepsilon'/(\varepsilon'^2 + \varepsilon''^2) + i\varepsilon''/(\varepsilon'^2 + \varepsilon''^2) \qquad (3.26)$$

ε' is the relative permittivity or dielectric constant and is calculated via

$\varepsilon' = C/C_0$, where C and C_0 are the capacitances of the full and empty conductance cell; ε'' is the dielectric loss and is given by $\varepsilon'' = \kappa(\omega)/\omega e_0$, where $\kappa(\omega)$ is the conductivity of the sample at frequency ω and e_0 the permittivity of free space. In the limit of high frequencies

$$\lim_{\omega\tau_\kappa \gg 1} \varepsilon' = \varepsilon_s = \frac{1}{M_s}$$

and for low frequencies

$$\lim_{\omega\tau_\kappa \ll 1} \varepsilon' = \varepsilon_s \frac{\langle\tau_\kappa^2\rangle}{\langle\tau_\kappa\rangle^2} = \frac{1}{M_s}\frac{\langle\tau_\kappa^2\rangle}{\langle\tau_\kappa\rangle^2} \qquad (3.27)$$

and

$$\lim_{\omega\tau_\kappa \ll 1} \kappa(\omega) = \kappa = \frac{e_0\varepsilon_s}{\langle\tau_\kappa\rangle} = \frac{e_0}{M_s\langle\tau_\kappa\rangle} \qquad (3.28)$$

where $\langle\tau_\kappa\rangle$ is the mean conductivity relaxation time.

The data for a 40-mol% $Ca(NO_3)_2$/60-mol% $K(NO_3)_2$ solution have been analyzed in terms of the above formalism using the decay function

$$\phi(t) = \exp\left[-(t/\tau_0)^\beta\right] \qquad 0 < \beta \leqslant 1 \qquad (3.29)$$

where τ_0 is the characteristic relaxation time and β a parameter inversely proportional to the width of the distribution of relaxation times. A selection of the data for β M_s, τ_0, $\langle\tau_\kappa\rangle$ are displayed in Table 3-2 for both liquid and glass; $\langle\tau_\kappa\rangle$ was calculated from

$$\langle\tau_\kappa\rangle = (\tau_0/\beta)\Gamma(1/\beta)$$

where Γ is the gamma function. The β parameter increases with decreasing temperature for the liquid but remains constant for the glass. This implies that the range of conductivity relaxation times in the liquid increases with increasing temperature. On the basis that the conductivity mechanism involves an ionic "hop" from one site of local free energy minimum to another and covers a distance of a few ionic diameters, an increase in conductivity relaxation times can be interpreted as an increase in structural microheterogeneity in the liquid. According to Angell[44] the nondirectional coulombic bonding present in ionic liquids may, at temperatures approaching T_g, produce a highly ordered random-close-packed structure; this would be destroyed with increasing temperature and decreasing density and produce greater fluctuations in the microstructure. However, if

TABLE 3-2. Electrical Field Relaxational Parameters
for $0.4Ca(NO_3)_2$–$0.6KNO_3$ Melt and Glass[a, b]

$T°C$	β	M_s	τ_0/s	$\langle \tau_\kappa \rangle /s$
Glass				
25.3	0.74	0.138	2.05	2.47
44.0	0.74	0.138	1.71×10^{-1}	2.06×10^{-1}
54.3	0.74	0.138	5.6×10^{-1}	6.7×10^{-2}
Liquid				
60.1	0.72	0.133	1.22×10^{-2}	1.5×10^{-2}
71.4	0.64	0.124	3.3×10^{-4}	4.6×10^{-4}
80.7	0.58	0.116	1.4×10^{-5}	2.2×10^{-5}
93.2	0.52	0.112	2.92×10^{-5}	5.5×10^{-7}

[a] Data from Ref. 43.   [b] $T_g = 59.5 °C$.

this model is correct, then in order to explain the constant β values at temperatures $> (T_g + 25$ K) it must be assumed that the liquid structure becomes significantly less temperature dependent or the conductance mechanism becomes insensitive to structure. A similar observation has been made for $Ca(NO_3)_2/6.15H_2O$ where at $(T_g + 34$–44 K) the distribution of relaxation times is constant.[45]

Both $Ca(NO_3)_2/6.15H_2O$ and $Ca(NO_3)_2/KNO_3$ liquids display non-symmetric conductivity relaxation spectra, and a symmetric distribution function of relaxation times, equation (3.29), fits on the low-frequency side of the distribution only and not on the high-frequency side. The reason for this is uncertain; however, the derived parameters M_s, and especially $\langle \tau_\kappa \rangle$, reproduce the direct current conductivity κ, equation (3.28), within acceptable error, indicative of the fact that longer relaxation times make the greatest contribution to the conductivity.

Shown in Fig. 3-10 are the data for shear and conductivity relaxation times in $0.4Ca(NO_3)_2/0.6KNO_3$ as a function of temperature. The shear relaxation time $\langle \tau_s \rangle$ is the average time constant for decay of shear stress in a liquid at constant strain via the process of viscous flow. At higher temperatures $T/T_g > 1.27$, $\langle \tau_\kappa \rangle \doteq \langle \tau_s \rangle$ so that charge transport is directly related to the process of viscous flow via structural rearrangement of local liquid structures, as predicted by the Walden product equation (1.34). At lower temperatures $\langle \tau_s \rangle > \langle \tau_\kappa \rangle$ and at T_g the ratio $\langle \tau_s \rangle / \langle \tau_\kappa \rangle \doteq 10^4$. Comparable results have been obtained from $Ca(NO_3)_2 \cdot 8H_2O$; however, quite different results have been obtained from other liquids, e.g., in the alkali network oxides $\langle \tau_s \rangle / \langle \tau_\kappa \rangle \doteq 10^{11}$ at T_g, and for $LiCl \cdot (4.83$–$5.77)H_2O$ the ratio is 5–10.[46] In the alkali network oxides it is to be expected that

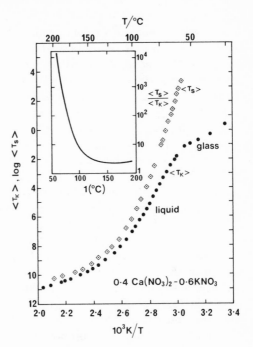

FIG. 3-10. Arrhenius plots of mean conductivity relaxation times $\langle \tau_\kappa \rangle$ and mean shear relaxation times $\langle \tau_s \rangle$ for 0.4Ca(NO$_3$)$_2$/0.6KNO$_3$. Insert shows $\langle \tau_s \rangle / \langle \tau_\kappa \rangle$ ratio versus temperature. Reprinted with permission from F. S. Howell, R. A. Bose, P. B. Macedo, and C. T. Moynihan, *J. Phys. Chem.* **78** 639–648 (Fig. 8) (1974). Copyright 1974 American Chemical Society.

the conductivity and hence $\langle \tau_\kappa \rangle$ would be related to the mobility of the mobile alkali metal ions, while the viscosity and hence $\langle \tau_s \rangle$ would be determined by the relaxation time of the network oxide structure resulting in a large $\langle \tau_s \rangle / \langle \tau_\kappa \rangle$ ratio near T_g. Similar processes must be occurring in the other melts mentioned above, but the disparity in relaxation times would be less because of the absence of a network structure. The 0.4Ca(NO$_3$)$_2$/0.16KNO$_3$ melt is in the range intermediate between silicate and LiCl/H$_2$O liquids, and this may be explained by assuming that near T_g the Ca^{2+} and NO$_3^-$ form structural elements whose relaxation times are long compared to those of the diffusional motions of the K$^+$ ions. LiCl/H$_2$O mixtures are distinct because of their low $\langle \tau_s \rangle / \langle \tau_\kappa \rangle$ ratios near T_g, and it has been suggested that this is due to the fact that both ions are singly charged and similarly bound into the low-temperature liquid structure. Thus shear and electrical relaxation times would be comparable because both ions would have a similar mobility, and conductivity relaxation would require cooperative rearrangement of most of the local microstructure.

An important feature of Fig. 3-10 is that of a reoccurrence of Arrhenius temperature dependence at low temperatures, $\log\langle \tau_\kappa \rangle \propto 1/T$. This feature has been observed of the shear viscosity of several glass-forming liquids[47] and indicates that the VTF equation is valid over a limited range only. A further example is given by the study of $Ca(NO_3)_2 \cdot xH_2O$, $(x = 4, 6, 8, 10)$[45] over an extended temperature range of 25 to $-83.3°C$ and illustrates that in order to fit the VTF equation to the experimental conductance data, T_0 must decrease with temperature. From the VTF equation (3.12), it can be seen that at low temperatures as the best-fit value of $T_0 \rightarrow 0$, equation (3.12) tends to the Arrhenius equation. For $Ca(NO_3)_2 \cdot xH_2O$ mixtures, the tendency to Arrhenius behavior occurs at higher temperatures, relative to T_0, for the most concentrated solutions.

The low-temperature Arrhenius region may be explained by postulating the onset of an ionic solidlike conductance mechanism as the liquid becomes more rigid at high $\langle \tau_s \rangle / \langle \tau_\kappa \rangle$ ratios.[33, 43] Thus there may occur a process where fluctuations in configurational entropy determine the relaxation time as well as a process where individual ions undergo successive displacement in a semirigid lattice. Both processes contribute to the conductance and hence to the slope of the $\ln\langle \tau_\kappa \rangle$ versus $1/T$ line, i.e., the activation energy. Figure 3-10 also shows that below the glass transition point the $\log\langle \tau_\kappa \rangle$ versus $1/T$ Arrhenius plot undergoes an abrupt change in slope, indicating a sudden drop in the activation energy for electrical conductance. Howell *et al.* propose that this is due to the dominance of an ionic solid type of conductance mechanism in the glass. Below T_g the glass structural relaxation times are so long compared to the time scale of conductance measurement that local structural rearrangements cannot contribute to the conductance mechanism and, therefore, the glass activation energy is correspondingly smaller than that for the liquid. It is interesting that the shape of the $E_{\Lambda, p}$ versus temperature curve for this melt follows the pattern predicted for $E_{\Lambda, V}$ versus volume predicted by Barton and Speedy,[39] i.e., that $E_{\Lambda, V}$ passes through a maximum with decreasing volume.

Electrical Conductivity in Ionic Liquids at High Temperatures

This chapter will be concerned with the discussion of ionic conductivity in ionized molten salts at temperatures well above T_0. At very high temperatures and low densities the fate of most ionic liquids is to become molecular, with a consequent drop in conductivity. The discussion of this phenomenon and of melts that are extensively molecular even at their melting points will be delayed to the next chapter. An ionic melt possesses a high number of charge carriers per unit volume, and if these are mobile the conductivity will be high. Klemm has defined an ionic melt as one whose conductivity $\kappa > 10^{-3}\,\text{S}\,\text{cm}^{-1}$. This definition is quite arbitrary, but it does seem to include most melts that are regarded as being predominantly ionized. A notable exception, for example, is the tetraalkyl ammonium tetrafluoroborate salts.

Electrical conductivity in ionic melts has been reviewed by many authors; see, for example, chapters in books by Blander,[1] Sundheim,[2] Mamontov,[3] Copeland,[4] Petrucci[5] and review articles by Janz et al.,[6, 7] Inman et al.,[8, 9] Tomlinson[10] and Tödheide.[11] In particular, the reader is referred to the chapter by Moynihan in *Ionic Interactions*.[5]

4.1. The Temperature and Pressure Dependence of Electrical Conductivity in Ionic Liquids

The temperature dependence, at constant pressure, of the equivalent conductivity of a selection of ionic melts is shown in Fig. 4-1. The approximate linearity of these Arrhenius plots results in an approximately constant activation energy, $E_{\Lambda,p}$, and these are displayed in Table 4-1. At lower temperatures it is expected that the conductivity isobars in Fig. 4-1 would show a substantial curvature toward the ordinate as the temperature fell below $2T_0$ (Chapter 3); low-temperature measurements, however, are

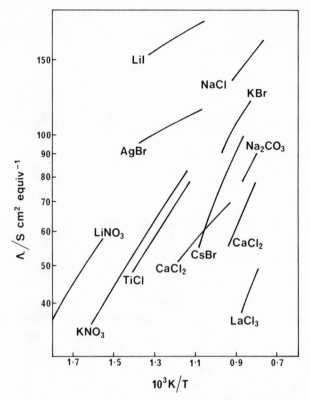

FIG. 4-1. Conductivity isobars for a range of ionic melts.[12]

usually impossible in one-component melts because of the untimely intervention of crystallization. At very much higher temperatures the conductivity isobars would pass through a maximum due to ion association, and this phenomenon is covered in Chapter 5. In the intermediate temperature range, Fig. 4-1, most ionic melts follows an Arrhenius law

$$\Lambda = A \exp(-E_\Lambda / RT)$$

and much attention has been given to the correlation of $E_{\Lambda,p}$ with ion size, shape, etc. A few trends are observable from Fig. 4-1 and Table 4-1. For the alkali halides, $E_{\Lambda,p}$ is the smallest and Λ the largest when $r_+ < r_-$; and $E_{\Lambda,p}$ increases and Λ decreases as $r_+ \to r_-$. The Ag halides do not follow this trend, the activation energies being much smaller than those of cations of comparable size, e.g., K^+. $E_{\Lambda,p}$ for the alkali nitrates seems to be independent of r_+ , although Λ decreases from Li^+ to Cs^+. Salts containing higher valence ions generally have higher activation energies and lower

TABLE 4-1. Conductivity Parameters for a Selection of Ionic Melts[a,b]

Cation	Cl⁻ ($r_- = 1.81$ Å)				Br⁻ ($r_- = 1.96$ Å)				I⁻ ($r_- = 2.20$ Å)			
	$E_{\Lambda,p}$	$E_{\Lambda,V}$	ΔV_Λ	$E_{\Lambda,V}/E_{\Lambda,p}$	$E_{\Lambda,p}$	$E_{\Lambda,V}$	ΔV_Λ	$E_{\Lambda,V}/E_{\Lambda,p}$	$E_{\Lambda,p}$	$E_{\Lambda,V}$	ΔV_Λ	$E_{\Lambda,V}/E_{\Lambda,p}$
Li⁺ ($r_+ = 0.68$ Å)	8.43 (916 K)	8.56	−0.8	1.01	8.86; 9.85 (869 K)	9.06	−0.3	1.02	7.57 (798 K)	7.57	~0.0	1
Na⁺ ($r_+ = 0.97$ Å)	10.59; 9.3 (1098 K)	5.96	3.0	0.56	9.92; 9.8 (1015 K)	6.41	0.65	0.65	9.63; 9.2 (979 K)	5.83	4.8	0.6
K⁺ ($r_+ = 1.33$ Å)	14.11; 14.0 (1065 K)	7.31	6.3	0.52	15.58; 15.1 (1015 K)	8.35	7.2	0.54	14.49; 14.5 (989 K)	7.35	9.7	0.51
Rb⁺ ($r_+ = 1.47$ Å)	15.68; 16.4 (1009 K)	8.0	8.0	0.52	15.50; 13.3 (988 K)	7.93	9.1	0.51	14.49; 15.4 (978 K)	6.73	11.4	0.46
Cs⁺ ($r_+ = 1.67$ Å)	16.74; 15.8 (944 K)	8.49	9.3	0.51	16.41; 15.5 (956 K)	8.88	10.6	0.54	15.89; 16.1 (952 K)	8.45	13.5	0.53
Ag⁺ ($r_+ = 1.26$ Å)	5.13; 4.7 (745 K)	2.78	1.7	0.54	4.62; 4.68 (727 K)	2.84	1.3	0.61				
Ca²⁺ ($r_+ = 0.99$ Å)	22,112				20,510				19,320			
La³⁺ ($r_+ = 1.14$ Å)	23,756				43,080							

continued overleaf

TABLE 4-1. (Continued)

	NO_2^-				NO_3^- ($r_- = 2.44$ Å)				CO_3^- ($r_- = 2.58$ Å)			
	$E_{\Lambda,p}$	$E_{\Lambda,v}$	ΔV_Λ	$E_{\Lambda,v}/E_{\Lambda,p}$	$E_{\Lambda,p}$	$E_{\Lambda,v}$	ΔV_Λ	$E_{\Lambda,v}/E_{\Lambda,p}$	$E_{\Lambda,p}$	$E_{\Lambda,v}$	ΔV_Λ	$E_{\Lambda,v}/E_{\Lambda,p}$
Li$^+$ $r_+ = 0.68$ Å					14.52 (400 K)	14.10	0.4	0.97	18.4			
Na$^+$ $r_+ = 0.97$ Å	12.6 (400 K)	8.4 (723 K)	3.7	0.67	12.97 (400 K)	8.24	3.8	0.64	17.5			
K$^+$ $r_+ = 1.33$ Å	13.0 (450 K)	5.4 (723 K)	5.4	0.42	15.02 (400 K)	6.74	7.0	0.45	19.5			
Rb$^+$ $r_+ = 1.47$ Å					16.40 (400 K)	8.16	7.8	0.50				
Cs$^+$ $r_+ = 1.67$ Å					15.31 (400 K)	6.94	8.9	0.45				
Ag$^+$ $r_+ = 1.67$ Å												

[a] Data from Refs. 12–15.
[b] Units are as follows: $E_{\Lambda,p}$ and $E_{\Lambda,v}$ in kJ mol^{-1}, ΔV_Λ in cm^3 mol^{-1}, and r in Å.

conductivities than 1 : 1 salts. Beyond these qualitative observations it is difficult to correlate the conductivities of a wide range of molten salts because of the poorly understood contribution of ionic polarizability, ionic association, ion shape and charge, and the effect of volume on the transport mechanism. Furthermore, the comparison of ionic conductivities and activation energies should be made at equivalent T/T_0 ratios but, as mentioned above, these are generally unknown for ionic melts.

A feature of recent conductance studies on ionic melts is that they have been carried out under pressure,[13-20] and it is now possible to consider the separate effect of temperature and volume. Table 4-1 also displays the 1-atm values of $E_{\Lambda,V}$ and ΔV_Λ; $E_{\Lambda,V}$ is obtained from the measured $E_{\Lambda,P}$, ΔV_Λ, α and β via equation (4.9). The data in this table illustrate that the volume dependence of ionic conductance in molten salts is significantly dependent on ion size. For example, electrical conductance in the lithium halides and nitrates is volume independent because $\Delta V_\Lambda \fallingdotseq 0$ and therefore $E_{\Lambda,V}/E_{\Lambda,P} \fallingdotseq 1$. ΔV_Λ increases as either the cation or anion radius is increased, but the ratio of activation energies seems to become reasonably constant at $r_+/r_- > 0.54$. At constant pressure, over the limited experimental temperature range, $E_{\Lambda,P}$ and $E_{\Lambda,V}$ are constant; ΔV_Λ, however, decreases with increasing temperature, Table 4-2. Very high-pressure studies show that for the alkali nitrates, $E_{\Lambda,P}$ and $E_{\Lambda,V}$ increase with increasing pressure (decreasing volume); but for the Na^+, K^+, and Rb^+ salts, ΔV_Λ decreases with decreasing volume, Tables 4-3 and 4-4. However, this volume dependence is only apparent above 20 kbar, Fig. 4-2, and careful and accurate studies below this pressure up to 1 and 5 kbar show

TABLE 4-2. Activation Volumes for Molar Conductivity in Alkali Nitrate Melts as a Function of Temperature at 1 bar Pressure[17, 18]

$T/°C$	$\Delta V/cm^3\,mol^{-1}$				
	$LiNO_3$	$NaNO_3$	KNO_3	$RbNO_3$	$CsNO_3$
275	0.9	—	—	7.6	—
300	0.7	—	—	7.5	—
325	0.6	4.1	—	7.3	—
350	0.6	4.0	6.6	7.2	—
375	0.5	3.8	6.3	7.0	—
400	—	3.7	6.3	7.0	—
425	—	3.6	6.4	—	8.2
450	—	3.5	6.4	—	8.3
475	—	—	—	—	8.3
500	—	—	—	—	7.9
550	—	—	—	—	7.4
600	—	—	—	—	7.2

TABLE 4-3. Molar Conductivity Parameters for $LiNO_3$
as a Function of Pressure and Volume[a]

P/kbar	$V/\text{cm}^3\,\text{mol}^{-1}$	$E_{\Lambda,p}/\text{kJ}\,\text{mol}^{-1}$	$E_{\Lambda,V}/\text{kJ}\,\text{mol}^{-1}$	$\Delta V_\Lambda/\text{cm}^3\,\text{mol}^{-1}$ (at 600°C)
0	$(44.9)^b$	15	—	≈ 0
5	(40.3)	—	—	≈ 0
	40	—	14	—
10	(37.2)	—	—	1.4
	36	—	17	—
15	(34.8)	—	—	1.4
20	(32.8)	21	—	1.4
	32	—	19	—
25	(31.2)	—	—	—
	30	—	20	—
30	(29.8)	24	—	1.4

[a] $E_{\Lambda,p}$ and $E_{\Lambda,V}$ are constant over the experimental range of 400–700°C.
[b] The volumes in parentheses are at 700°C and are therefore the maximum volumes at any pressure, the maximum pressure occurring at constant volume over a range of temperatures can also be interpolated from the data.[20]

that $E_{\Lambda,V}$ and ΔV_Λ are independent of pressure.[17, 14] The general trend toward smaller ΔV_Λ's at higher pressures for $NaNO_3$, KNO_3, and $RbNO_3$ melts is quite the opposite of that observed in low-temperature molten salts where ΔV_Λ increases with increasing pressure. However, the increase in $E_{\Lambda,V}$ with increasing pressure (decreasing volume) is observed at lower temperatures for melts near T_0. The tetraalkylammonium tetrafluoroborate

TABLE 4-4. Molar Conductivity Parameters for $NaNO_3$
as a Function of Pressure and Volume[a]

P/kbar	$V/\text{cm}^3\,\text{mol}^{-1}$	$E_{\Lambda,p}/\text{kJ}\,\text{mol}^{-1}$	$E_{\Lambda,V}/\text{kJ}\,\text{mol}^{-1}$	$\Delta V_\Lambda/\text{cm}^3\,\text{mol}^{-1}$ (at 600°C)
0	(51.7)	14	—	3.7
	50	—	8.5	—
5	(45.9)	—	—	3.6
10	(42.5)	15	—	3.5
	40	—	—	—
15	(40)	—	—	3.4
20	(38.1)	17	—	3.2
30	(35.3)	19	—	2.4
	35	—	—	—
40	(33.1)	21	—	2.0
50	(31.4)	23	—	1.8
	30	—	17	—

[a] For explanation see footnotes to Table 4-3.

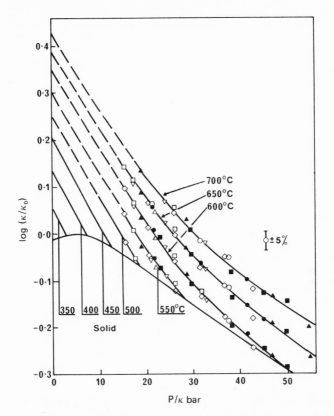

FIG. 4-2. Log conductivity of $NaNO_3$ versus pressure at constant temperature, $\kappa_0 = 1\ \text{S cm}^{-1}$. Reprinted with permission from V. Pilz and K. Tödheide, *Ber. Bunsenges. Phys. Chem.* **77**, 29–36 (Fig. 4) (1973).

melts show the same trends at low pressures up to 2 kbar.[19] $E_{\Lambda,V}$ increases with decreasing volume and decreasing temperature (Fig. 4-3), but, because of the fact that $r_+ > r_-$, $E_{\Lambda,V}$ shows very little dependence on cation volume in the series from tetrapropylammonium to tetraoctylammonium tetrafluoroborate. ΔV_{Λ} is also relatively independent of cationic radius and for a given salt decreases with increasing pressure.

The above is a brief but concise description of the effects of temperature and pressure on the electrical conductivity in one-component ionic melts at high temperatures. The quantitative prediction of these effects has proved to be very difficult, but various models can at least qualitatively account for some of the observed trends. What follows below is a description of the theoretical treatment of conductance in molten salts with emphasis on the development of statistical theories.

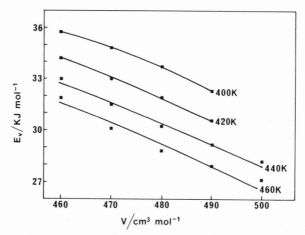

FIG. 4-3. Plot of E_V against volume at constant temperature for n-(hexyl)$_4$NBF$_4$. Note that an equivalent figure in Ref. 19 is incorrectly labeled. Reprinted with permission from A. F. M. Barton and R. J. Speedy, *J. Chem. Soc. Faraday Trans. I*, **70**, 506–527 (Fig. 6) (1974).

4.2. Theories for Electrical Conductivity in Ionic Melts

4.2.1. Transition State Theory

The form of this theory, originally due to Glasstone, Laidler, and Eyring,[21] that is most preferred results from the treatment by Bockris *et al.*[22] Their treatment was originally derived for the case of silicate melts but has been applied to molten salts, molten salt mixtures, and electrolyte solutions (Chapter 2).

The transition state for the process of mass transport in liquids is defined with reference to a quasilattice structure, Fig. 4-4, where the surrounding ions are regarded as being stationary. Before the application of the electric field E, motion is random and nondirectional, and the free-energy barrier opposing translation is ΔG_i. When the electric field is applied, however, motion becomes biased in the direction of the electric field because the field lowers the activation energy by an amount $\delta G_i^{\ddagger} = Ez_i e d_i \cos \phi$, where $z_i e$ is the charge on the ion i. The velocity v_i, of an ion i, in the direction of the applied field is given by $v_i =$ (distance traveled) $\times$ (rate of barrier crossing) or

$$v_i = 2d_i \cos \phi \left\{ \frac{kT}{h} \exp\left[-\frac{(\Delta G_i^{\ddagger} - \delta G_i^{\ddagger})}{RT} \right] \right\} \tag{4.1}$$

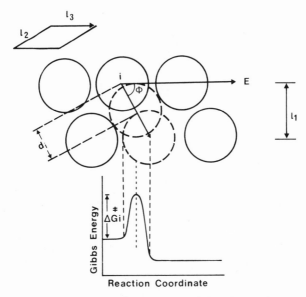

FIG. 4-4. Transition state diagram for the calculation of transport parameters.

where the term in the curly braces arises directly from transition state theory. After averaging for motion over all values of ϕ, the resultant velocity $\bar{v}_i$ in the direction of the applied field is

$$\bar{v}_i = 2d_i \frac{kT}{h} \exp\left(\frac{-\Delta G_i^{\ddagger}}{RT} \right) \frac{a}{3} \tag{4.2}$$

where $a = Ez_i Fd_i / RT \ll 1$.
The molar conductivity is then given by

$$\lambda_i = |z_i| \bar{v}_i F = \frac{2}{3} |z_i| d_i Fa \frac{kT}{h} \exp\left(\frac{-\Delta G_i^{\ddagger}}{RT} \right) \tag{4.3}$$

E, the effective potential gradient acting on an ion is difficult to evaluate, but if it is written as $E = (\varepsilon + 2)/3$, where ε is the relative permittivity, then

$$\Lambda = \sum_i \lambda_i = \sum_i 3.62 \times 10^{19} z_i^2 d_i^2 \exp\left(-\Delta G_i^{\ddagger}/RT \right) \tag{4.4}$$

and

$$\Delta G_i^{\ddagger} = \Delta H_i^{\ddagger} - T\Delta S_i^{\ddagger} \tag{4.5}$$

Equations similar to those of (2.46)–(2.49) result, i.e.,

$$E_{\Lambda,P} = E_P = RT^2 \left(\frac{\partial \ln \Lambda}{\partial T} \right)_P = \Delta H^{\ddagger} + \frac{2}{3} \alpha RT \qquad (4.6)$$

$$E_{\Lambda,V} = E_V = RT^2 \left(\frac{\partial \ln \Lambda}{\partial T} \right)_V = \Delta U^{\ddagger} \qquad (4.7)$$

and

$$\Delta V_{\Lambda} = -RT \left(\frac{\partial \ln \Lambda}{\partial P} \right)_T = \Delta V^{\ddagger} \qquad (4.8)$$

where

$$E_P = E_V + (\pi_I + P)\Delta V_{\Lambda} \qquad (4.9)$$

At moderate pressures and over a limited temperature range, equation (4.4) provides a qualitative explanation of the linear $\log \Lambda$ versus $1/T$ plots in Fig. 4-1. The slope of these plots E_P, equation (4.6) cannot, however, be predicted from the theory. From equation (4.9) E_P is seen to be the sum of two parts, the constant volume activation energy E_V and the expansion work for the process (volume of initial state, V_I) → (volume of activated state $V^{\ddagger}$). Now $\Delta V^{\ddagger} = V^{\ddagger} - V_I$ and in an ionic molten salt where $r_+ < r_-$, e.g., LiCl, the smaller ion would be the most mobile and $V_I \doteq V^{\ddagger}$, $\Delta V \doteq 0$ as observed. On this basis ΔV would be expected to increase as $r_+ \rightarrow r_-$, as observed. In fact, ΔV increases with ion size as either r_+ or r_- is increased, Table 4-1. The theory correctly predicts that in the case where $r_+ \ll r_-$, $\Delta V \doteq 0$ and $E_P \doteq E_V$, and that when $r_+ \rightarrow r_-$, $E_P > E_V$. $E_V = \Delta U^{\ddagger}$ is the work required to form the activated state at constant volume and could be calculated from a detailed knowledge of the structure of the initial and activated states and the anion–cation pair potential. However, such calculations have not been performed (to the writer's knowledge), and in any case $E_{\Lambda,V}$ is approximately constant for the alkali halides, nitrates, and nitrites and shows no systematic variation with ion size or polarizability.

$E_{\Lambda,P}$ increases for salts containing multivalent ions, and there has been temptation to relate this to enhanced anion–cation interaction. Presumably this could arise from an increase in E_V and π_I (the internal pressure) in equation (4.9). However, while the long-range coulomb potential does affect the equilibrium properties of ionic melts, it appears to have an insignificant role in determining the transport coefficients (see below) and therefore should have little effect on E_V. High-pressure conductance studies might provide some explanation of how E_P is enhanced by multivalency. Transition state theory has been criticized on several bases as being inapplicable to high-temperature ionic melts. These criticisms are based on the requirement that the local molecular environment containing the activated molecule be fixed during the jump from one equilibrium site to another and that the jump itself be clearly distinguishable from the "rattling" motion executed by an ion in its cage of nearest neighbors.

The first requirement, that the lifetime of the local molecular structure that comprises the transition state be significantly longer than the time required for a molecule to jump from one site to another, has been tested by Goldstein.[23] He has estimated the longest possible time, τ, a local region of dimension $l = 10^{-7}$ cm has to complete the transition process once it has become activated. This is the time required for energy to flow out of the region and is assumed to be determined by the thermal diffusivity κ_D, so that $\tau = l^2 / \kappa_D$. A typical value of κ_D for liquids is 10^{-3} cm^2 s^{-1}; thus $\tau \doteq 10^{-11}$ s. In order for the above constraint to be obeyed, the relaxation time of the local liquid structure would need to be about 10^{-9} s which, if regarded as the shear viscosity relaxation time, corresponds to a viscosity of ~ 10 poise. This is about 10^3 times the viscosity of high-temperature ionic melts and indicates that transition state model is not applicable to these liquids. Furthermore, the experimental activation energies are not significantly greater than the mean kinetic energy of an ion in an ionic melt; $E_{\Lambda, P}$ varies from 0.7 to 3 RT, which suggest that a significant fraction of ions can at any instant be in an activated state, and the concept of discrete ion jumps resulting from energy activation becomes inappropriate.

Computer simulation of a hard-sphere fluid has shown that the mean free path distribution decreases almost exponentially with increasing distance and does not exhibit a peak near a free path distance equal to twice the ionic radius.[24] This distance is regarded as the most probable one if a jump model is obeyed and the calculated probability of a jump of this distance is not large enough to account for experimental diffusion coefficients. More recent computer simulation experiments for molten salts and molecular liquids provide further evidence contrary to the requirements of a jump model. In molten KCl at 1045 K, for example, the velocity autocorrelation function becomes zero at about 30×10^{-14} s,[25] and if a discrete jump model of ionic mobility were valid the force autocorrelation function should become zero in, at the most, one-tenth of this time. However, it too becomes zero at about 30×10^{-14} s. This suggests that a particle is under constant acceleration, and a discrete jump model is not valid.

4.2.2. The Hole Model of Liquids

This theory, due to Fürth,[26] assumes that on melting the appearance of ion-sized holes destroys the long-range order in a solid and it becomes a fluid. Charge transport occurs when, under the influence of an externally applied potential gradient, an ion jumps into an adjacent hole. Thus the barriers to ionic migration are ΔH_h, the enthalpy of hole formation, and $\Delta H_j^{\ddagger}$, the energy required for the jumping process. From the theory of activated rate processes, Bockris *et al.*[27] derived the following equation for conductivity in ionic melts

$$\Lambda = \frac{A}{T}\exp\left(-\frac{\Delta H_h + \Delta H_j^{\ddagger}}{RT}\right) \qquad (4.10)$$

and from equations (4.6) and (4.10), and by assuming that $\Delta H_h \gg \Delta H_j^{\ddagger} \simeq 0$ then $\Delta H_h = E_{\Lambda,P} + RT$. At constant volume a similar analysis shows that $\Delta H_j^{\ddagger} = E_{\Lambda,V} + RT$ or $E_{\Lambda,V} \doteq -RT$. Now at low pressures the Gibbs energy of hole formation is given by $\Delta G_h = 4\pi r^2 \sigma(T)N_A$, where r is the hole radius and $\sigma(T)$ the temperature-dependent macroscopic surface tension. By comparison with the thermodynamic identity $\Delta G = \Delta H - T\Delta S$, the above equation gives $\Delta H_h = 4\pi r^2[\sigma - \partial\sigma/\partial T(T - T_m)]N_A$. But this did not allow for the temperature dependence of r, which is related by Fürth to the average hole volume by

$$\overline{V}_h = \frac{4}{3}\pi\bar{r}^3 = 0.68\left(\frac{kT}{\sigma(T)}\right)^{3/2} \qquad (4.11)$$

When this equation is taken into account, $\Delta H_h = 0$. This prediction is clearly in error since $E_{\Lambda,P}$ varies significantly from RT. A more recent treatment of the theory due to Emi and Bockris[28] predicts that the activation energy for diffusion $E_{D,P} \doteq 3.5RT_m$. They show that with considerable scatter, there is an approximate relationship between the experimental $E_{D,P}$ energies and T_m, the melting temperature. This relationship must be empirical since the temperature of the solid–liquid transition cannot be regarded as the reference point for transport coefficients in liquids whose temperature range is bounded by the glass transition temperature and the critical temperature.

Activation volumes for conductance in molten salts vary widely and bear little relation to the value predicted by equation (4.11) namely, the volume of a mole of holes or approximately the molar volume. Furthermore, the ratio of constant volume activation energies $E_{\Lambda,V}$ to the constant pressure activation energies $E_{\Lambda,V}/E_{\Lambda,P}$ is for some liquids quite large (Table 4-1), in contradiction to the assumption made above that $\Delta H_h \gg \Delta H_j^{\ddagger}$. For the reasons given here and above (Section 4.2.1.) the theory cannot be regarded as a realistic description of transport in ionic melts nor of liquids in general.

4.2.3. The Theory of Significant Liquid Structures

The theory of significant liquid structures has been described in detail in several publications and will not be developed in detail here.[29, 30] However, because of its unique ability to account for the thermodynamic and transport properties of liquids and the extent to which the theory has been tested against experimental data, it is incorporated into this book.

The basis of the theory is the postulate that a liquid contains fractions of solidlike and gaslike particles. Regions of regular close-packed structure form the solidlike structure and the gaslike fraction arises from fluidized vacancies randomly interspersed throughout the liquid. The vacancies are thought to move through the liquid as freely as a gas molecule in a gas and to confer gaslike properties to the surrounding molecules equivalent to one gas molecule. For N molecules the total partition function can thus be written as the weighted product of the partition function for solidlike f_s and gaslike f_g molecules:

$$f^N = f_s^{N(V_s/V)} f_g^{N[(V-V_s)/V]} \tag{4.12}$$

V_S is the volume of the solid; and, at a volume V, the ratio V_s/V is the fraction of solidlike molecules, and the ratio $(V - V_s)/V$ is the fraction of gaslike molecules. Since f_g is known and an approximate form of f_s can be derived, then f^N, and hence the thermodynamic properties of the liquid, can also be calculated.

In an analogous manner the viscosity of a liquid can be regarded as being a weighted sum of the viscosities of solidlike and gaslike ions

$$\eta = (V_s/V)\eta_s + (V - V_s)/V\eta_g \tag{4.13}$$

Viscosity is defined by the ratio of the shear stress to the shear rate, and the rate of shear in the solidlike fraction is calculated by assuming that momentum transport occurs through a series of activated jumps by a molecule in one layer to a hole in an adjacent layer. The analysis is similar to that presented in Section 4.2.1 and Fig. 4-4 for conductance but makes the assumption that the activation energy required for jumps is very much less than the kinetic energy of a molecule in the liquid. This would imply that at any instant a large fraction of molecules are "activated" and, as discussed above, under these conditions the concept of transition states and discrete jumps becomes invalid. Notwithstanding these objections the viscosity of the solidlike fraction is given by

$$\eta_s = \frac{kT}{\kappa'} \left/ \frac{(2d_i)^2 l_2 l_3}{2l_1} z \frac{V - V_s}{V} \frac{1}{3} \right. \tag{4.14}$$

κ' is the frequency of jumping into a neighboring empty position in the pseudo lattice and is given by transition state theory and significant structures theory as

$$\kappa' = k\frac{kT}{h}\left[1 - \exp(-\theta/T)\right]\exp\left(-\frac{a'E_sV_s}{(V - V_s)RT}\right) \qquad (4.15)$$

θ is the Einstein characteristic temperature of a one-dimensional oscillation, E_s the sublimation energy of the solid, a' a constant, and z the coordination number. For the purpose of calculation it is assumed that $2d_i = l_1 = l_2 = l_3$ and

$$l^3 = 2^{1/2}V_s/N \qquad (4.16)$$

The gaslike contribution is derived from the kinetic theory of gases as

$$\eta_g = \frac{2}{3d^2}\left(\frac{mkT}{\pi^3}\right)^{1/2} \qquad (4.17)$$

where d is the molecular diameter. An expression for the viscosity of the liquid is obtained by combining equations (4.13)–(4.17). However, in the case of molten salts the gaslike phase is presumed to consist of neutral monomeric and dimeric molecules, the fractions of which are calculated from significant structures theory as applied to equilibrium properties. Thus equation (4.17) becomes a composite sum of contributions from both types of molecules.

The diffusion coefficient can be calculated from the total viscosity equation and the Stokes–Einstein relation and the conductivity from the Nernst–Einstein relation.[31] The final result is

$$\kappa = \frac{e^2}{4\pi\eta}\left(\frac{V_s}{V}\right)\left(\frac{N}{V}\right)\left(\frac{z_+}{r_+} + \frac{z_-}{r_-}\right) \qquad (4.18)$$

where z_+, r_+, z_-, r_- are, respectively, charges and ionic radii of the cations and anions. The factor NV_sV^{-2} in equation (4.18) is the fraction of solidlike ions since only these ions are presumed to conduct because the gaslike fraction is assumed to consist of neutral molecules. In the calculation of the conductivity by equation (4.18), the values of κ' and a' were chosen for the best data fit over a range of ionic melts, the values for the other parameters being the same as those used in the calculation of thermodynamic properties. For NaCl, NaBr, KCl, KBr, and KI the calculated conductivities were 48 to 26% too low. Given the nature of the assumption made in the derivation of the theory, the fit must be regarded as good. Moynihan,[5] however, claims that the fit is primarily the result of curve fitting since the dominant terms in the viscosity equation vary little from salt to salt, and the two adjustable parameters make a good fit to equation (4.18) inevitable as the experimental data do not vary by more than 25%.

The principal weakness of the significant structures approach in its present form, when applied to transport properties, arises from the use of transition state theory to calculate the viscosity of the solidlike fraction of molecules. This weakness does not arise from the fundamental postulates of the theory and could conceivably be improved upon by an alternative, but as yet unknown, treatment.

4.2.4. The Kirkwood–Rice–Allnatt Kinetic Theory of Electrical Conductance

Statistical mechanical theories of liquid state equilibrium properties have advanced significantly over the last decade or so,[32] but little progress has been made in the statistical mechanical treatment of molten salts transport properties beyond that of Berne and Rice.[33]

The three fundamental problems in the development of a statistical mechanical theory of transport in liquids have been identified as:[34]

a. the means by which the time reversible equation of classical mechanics used to describe the motion of molecules lead to time irreversible transport equations;
b. the derivation of a suitable "kinetic equation" to determine the time evolution of some phase space ensemble probability distribution function;
c. solution of the "kinetic equation" to derive relationships between the transport coefficients and the molecular properties.

In accordance with the pioneering work of Kirkwood, Rice and Allnatt[36] have found an approximate solution to these problems for a one-component fluid.

If

$$f^{(N)} \equiv f^{(N)}(\mathbf{p}_1, \mathbf{p}_2, \ldots, \mathbf{p}_N; \mathbf{r}_1, \mathbf{r}_2, \ldots, \mathbf{r}_N) \, d\mathbf{p}_1, d\mathbf{p}_2, \ldots, d\mathbf{p}_N, d\mathbf{r}_1, d\mathbf{r}_2, \ldots, d\mathbf{r}_N \tag{4.19}$$

is the phase space probability distribution function of N particles, then its time evolution, according to the classical equations of motion, is given by the Liouville equation

$$-\frac{\partial f^{(N)}}{\partial t} = \left(\sum_{i=1}^{N} \frac{\mathbf{p}_i}{m} \cdot \nabla_i + \sum_{i=1}^{N} \mathbf{F}_i \cdot \nabla_{\mathbf{p}_i} \right) f^{(N)} \tag{4.20}$$

$\mathbf{p}_i$ is the momentum of particle i and $\mathbf{F}_i$ is the intermolecular force on i due to the other $(N-1)$ molecules. Irreversibility is introduced into this equation by defining the "time coarse-grained" distribution function $f^{(n)}$ as

$$\bar{f}^{(n)}(\Gamma_n ; t) = \frac{1}{\tau} \int_0^\tau f^{(n)}(\Gamma_{n,t} ; t+s)\, ds$$

where $\Gamma_{n,t}$ denotes the phase Γ_n of the set of n molecules at time t. Thus $\bar{f}^{(n)}$ is the average value of $f^{(n)}$ over the interval τ. The time τ is chosen so that the molecular events occurring in one interval of τ are independent of events in the preceding intervals. More specifically, Rice and Allnatt define these events as a hard-core collision followed by erratic or quasi-Brownian motion in the fluctuating soft force field of the neighboring molecules. In order for this event, defined in time τ, to be independent of preceding events, the fluctuating soft force field should have reduced all force and momentum correlation between molecules to zero before the next collision and interval τ begins. Successive collisions are uncorrelated; the system has no memory of past events and the process becomes irreversible.

For diffusion only the singlet distribution function $\bar{f}^{(1)}$ is required. Equation (4.20) now becomes

$$\left(\frac{\partial}{\partial t} + \frac{1}{m} \mathbf{p}_1 \cdot \nabla_1 \right) \bar{f}^{(1)} = \Omega_H + \Omega_S \tag{4.21}$$

where

$$\Omega_H = -\frac{1}{\tau} \int_0^\tau \int \mathbf{F}_{12}^{(H)} \cdot \nabla_{\mathbf{p}_1} f^{(2)}(\Gamma_{2,t} ; t+s)\, d\Gamma_1(2)\, ds$$

$$\Omega_S = -\frac{1}{\tau} \int_0^\tau \int \mathbf{F}_{12}^{(S)} \cdot \nabla_{\mathbf{p}_1} f^{(2)}(\Gamma_{2,t} ; t+s)\, d\Gamma_1(2)\, ds$$

$\mathbf{F}_i$ in equation (4.20) has been replaced by $\mathbf{F}_{ij}$ on the principle of pairwise superposition of intermolecular forces, and the force has been broken up into a hard repulsive force between particles 1 and 2, $\mathbf{F}_{12}^{(H)}$, and a soft or attractive force $\mathbf{F}_{12}^{(S)}$; $f^{(2)}$ is the doublet distribution function and is replaced during the solution of equation (4.21) by $\bar{f}^{(2)}$, this in turn is replaced by the approximation that

$$\bar{f}^{(2)}(\mathbf{p}_1, \mathbf{p}_2, \mathbf{r}_1, \mathbf{r}_2 ; t) = \bar{f}^{(1)}(\mathbf{p}_1, \mathbf{r}_1 ; t) \bar{f}^{(1)}(\mathbf{p}_2, \mathbf{r}_2 ; t) g^{(2)}(\mathbf{r}_1, \mathbf{r}_2)$$

where $g^{(2)}(\mathbf{r}_1, \mathbf{r}_2)$ is the pair correlation function. A further mathematical approximation made during the solution of equation (4.21) is the assumption that cross correlations between hard and soft forces are negligible. This arises from the physical assumptions made in defining the dynamic event where the relaxation time of the soft forces must necessarily be significantly less than those of the hard forces in order to reduce velocity correlation to zero before the next collision.

For the case of diffusion the "kinetic equation" that results from the solution of equation (4.21) is an integrodifferential equation describing the time evolution of $\bar{f}_i^{(1)}$, the coarse-grained singlet distribution function of foreign molecules, i in dilute solution of host molecules. If X_1 is the mean force acting on a foreign molecule arising from a concentration gradient, or an external electric field, then it can be shown from a solution of the kinetic equation that the mean velocity of molecules 1 in the direction of X_1 is[37]

$$\bar{v}_1 = \frac{X_1}{\zeta^H + \zeta^S} \qquad (4.22)$$

If X_1 arises from a gradient of chemical potential then

$$X_1 \doteq -\frac{kT}{c_1}\nabla_1 c_1 \qquad (4.23)$$

where c_1 is the concentration, and, from the definition of the diffusion coefficient of molecules 1 in host molecules 2,

$$c_1\bar{v}_1 = -D_{12}\nabla_1 c_1 \qquad (4.24)$$

Equations (4.22)–(4.24) give

$$D_{12} = \frac{kT}{\zeta^H + \zeta^S} \qquad (4.25)$$

where ζ^H and ζ^S are the hard and soft friction coefficients. The soft force friction coefficient can be evaluated as follows from Brownian motion theory. A Brownian particle small enough to respond to fluctuations in force over its surface undergoes an increment in momentum $\Delta \mathbf{p}$ averaged over a time τ given by

$$\langle \Delta \mathbf{p} \rangle = \frac{\zeta_0}{m}\mathbf{p}\tau$$

If it is now assumed that the instantaneous force on a particle can be separated into a frictional force $-(\zeta_0/m)\mathbf{p}$ and a fluctuating force $\mathbf{X}(t)$ then

$$\mathbf{F}(t) = \frac{d\mathbf{p}}{dt} = -\frac{\zeta_0}{m}\mathbf{p} + \mathbf{X}(t)$$

$\mathbf{X}(t)$ arises from the surrounding molecules and is independent of the motion of the reference particle. Rice and Gray[34] have shown that the autocorrelation function of the total force is given by this equation to be

$$\langle \mathbf{F}(t)\cdot\mathbf{F}(t+\tau)\rangle = 3\phi(\tau) - \frac{3\zeta_0^2 kT}{m}\exp\left(-\frac{\zeta_0 t}{m}\right)$$

where $\phi(\tau) = \langle\mathbf{X}(t)\cdot\mathbf{X}(t+\tau)\rangle$ and $\phi(\tau) = 0$ for $\tau > \tau_1$ where $\tau_1 \ll m/\zeta_0$. The integral $I(\tau)$ of $\langle\mathbf{F}(t)\cdot\mathbf{F}(t+\tau)\rangle$ over a range $0 < \tau_1 < \tau$ is

$$I(\tau) = 3\int_0^\tau \phi(\tau)\,d\tau + 3\zeta_0 kT\left[\exp\left(-\frac{\zeta_0\tau}{m}\right) - 1\right]$$

Providing the above conditions on $\phi(\tau)$ are satisfied, then the first term on the right-hand side reaches a plateau value of $3\zeta_0 kT$ when $\tau = \tau_1$, while the second term on the right is still small; therefore,

$$I(\tau_1) = 3\int_0^{\tau_1} \phi(\tau)\,d\tau + 3\zeta_0 kT\left[\exp\left(-\frac{\zeta_0\tau_1}{m}\right) - 1\right]$$
$$\doteq 3\zeta_0 kT$$

and

$$\zeta_0 = \frac{1}{3kT}\int_0^{\tau_1} \langle\mathbf{F}(t)\cdot\mathbf{F}(t+\tau)\rangle\,d\tau \qquad (4.26)$$

By analogy with equation (4.26) the soft force friction coefficient for particles of equal mass is given by

$$\zeta^S = \frac{1}{3kT}\int_0^{\tau_1} \langle\mathbf{F}_S(t)\cdot\mathbf{F}_S(t+\tau)\rangle\,d\tau \qquad (4.27)$$

This is an approximation, the extent of which will be determined by agreement between experiment and theory.

Other methods of calculation of ζ^S are available and will be discussed below.

The hard force friction coefficient has been derived by Helfand[38] to be

$$\zeta^H = 8/3\rho g(d) d^2(\pi mkT)^{1/2} \tag{4.28}$$

where d is the diameter of a particle, $g(d)$ is the contact pair correlation function, and ρ is the number density. This equation was derived for a macroscopic Brownian particle.

Molten salts are regarded as two-component fluids by statistical mechanical theories, and as such present another degree of complexity. Rice and Berne[33] in attempting to overcome this problem define an ideal ionic melt with the following properties:

1.   Oppositely charged ions are of nearly equal size and have identical electronic properties except for sign of charge.

2. On the average, a positive ion is surrounded by negative ions and vice versa.

3.   The probability of a hard collision between like ions is negligible.

4.   The total pair potential is the sum of a rigid core repulsion, a van der Waals'-type attraction, and a coulombic attraction or repulsion modulated by polarization of the surrounding fluid.

With the preceding assumptions they derive an integrodifferential equation using the basic methods of the Rice–Allnatt theory of liquids described above. There are two kinetic equations for the coarse-grained singlet distribution function, one for each ion, and two friction coefficients representing interactions between like and unlike ions. For anions α and cations β, the anion friction coefficient is

$$\zeta_\alpha = \zeta_{\alpha\beta} + \zeta_{\alpha\alpha}$$

and

$$\zeta_{\alpha\alpha} = \zeta_{\alpha\alpha}^H + \zeta_{\alpha\alpha}^S$$
$$\zeta_{\alpha\beta} = \zeta_{\alpha\beta}^H + \zeta_{\alpha\beta}^S \tag{4.29}$$

Similar expressions exist for cations, incorporating $\zeta_{\alpha\beta}$ and $\zeta_{\beta\beta}$.

The solution of the kinetic equation due to Berne and Rice, in conjunction with the Bearman and Kirkwood expression for momentum and energy flow in a binary mixture, has been used to derive expressions for the shear viscosity and thermal conductivity. The mobility of cations and anions in an ideal ionic melt have also been derived. If $\bar{v}_{\alpha 1}$ is the average drift velocity of an anion in the direction of the applied field E, then it can be shown that

$$\bar{v}_{\alpha 1} = \frac{e_\alpha E + (F^*_{\alpha 1})_x}{\zeta^H_{\alpha\beta} + \zeta^S_{\alpha\alpha} + \zeta^S_{\alpha\beta}} \qquad (4.30)$$

where α and β represent anions and cations, $(F^*_{\alpha 1})_x$ is the average force on a representative anion $\alpha 1$ in the direction of the electric field z. The force arises from the distortion of spherical symmetry due to the fact that an ion and its oppositely charged environment of other ions have a force of opposite sign exerted on them by the electric field. The mobility of the ion is then given as

$$u_{\alpha 1} = \frac{e_\alpha}{\zeta^H_{\alpha\beta} + \zeta^S_{\alpha\alpha} + \zeta^S_{\alpha\beta}} \left[1 - \Delta\right]$$

where

$$\Delta = \frac{4\pi\rho_\beta}{3kT} \int_{d_{\alpha\beta}}^{\infty} \frac{dU(r_{\alpha\beta})}{dr_{\alpha\beta}} g^2(r_{\alpha\beta}) r^3_{\alpha\beta}\, dr_{\alpha\beta} \qquad (4.31)$$

$d_{\alpha\beta}$ is the hard-core contact distance, ρ_β the number density of cations, $U(r)$ the pair potential, and $g^2(r_{\alpha\beta})$ the pair distribution function. From equation (4.25) it can be seen that (4.31) can be written as

$$u_{\alpha 1} = \frac{e_\alpha D_\alpha}{kT} \left[1 - \Delta\right] \qquad (4.32)$$

where D_α is the diffusion coefficient of anions. The molar conductivity can be derived from equation (4.32) as

$$\frac{\Lambda}{F^2} = \frac{z^2_\alpha D_\alpha + z^2_\beta D_\beta}{kT} \left[1 - \Delta\right] \qquad (4.33)$$

This predicts deviations from the Nernst–Einstein equation for which $\Delta = 0$.

Several attempts have been made to calculate transport properties of molten alkali halides from this theory. Morrison and Lind[39] attempted to calculate the electrical conductivity of molten KCl, NaI, and the self-diffusion coefficients of K^+ and Cl^- in KCl and of I^- in NaI. They used an approximate theory for diffusion in a pure one-component single-species liquid called the "small step diffusion model" derived by Rice and Kirkwood. However, their analysis is wrong because they should have used small step diffusion theory derived for liquid mixtures or two-component liquids. As a consequence their computed results do not compare at all well with experiment.

Ichikawa and Shimoji[40] have used the Rice and Allnatt theory for liquid mixtures to calculate the viscosity and self-diffusion coefficient of anions and cations in five alkali halides. The theory gives the soft friction coefficient in a mixture of α and β (anion and cation) molecules as

$$\zeta_\alpha^S = \zeta_{\alpha\alpha}^S + \zeta_{\alpha\beta}^S$$

where

$$\zeta_{\alpha\gamma}^S = K_{\alpha\gamma}\left[\frac{1}{\zeta_\alpha^S} + \frac{1}{\zeta_\gamma^S}\right] \tag{4.34}$$

and

$$K_{\alpha\gamma} = \left(\frac{\rho_\gamma\mu_{\alpha\gamma}}{3}\right)\int_{d_{\alpha\gamma}}^{\infty}\nabla^2 U(r_{\alpha\gamma})g^2(r_{\alpha\gamma})\,d^3r \tag{4.35}$$

$$\gamma = \alpha \text{ or } \beta$$

where $\mu_{\alpha\gamma}$ is the reduced mass of the pair $\alpha\gamma$. From these relations Ichikawa and Shimoji have derived the equation

$$\left(\zeta_\alpha^S\right)^2 = \left[K_{\alpha\beta} + 2K_{\alpha\alpha}\right]\left[1 + \frac{K_{\alpha\beta}}{\left(K_{\alpha\beta} + 2K_{\alpha\alpha}\right)^{1/2}\left(K_{\alpha\beta} + 2K_{\beta\beta}\right)^{1/2}}\right] \tag{4.36}$$

The hard-core friction coefficient is given by

$$\zeta_\alpha^H = \sum_{\gamma=\alpha,\beta} 8/3\, d_{\alpha\gamma}^2 g^2(d_{\alpha\gamma})\rho_\gamma(\pi\mu_{\alpha\gamma}kT)^{1/2} \tag{4.37}$$

In order to evaluate ζ^S and ζ^H, the pair potential was assumed to have the form

$$U(r_{\alpha\gamma}) = U^H(r_{\alpha\gamma}) + U^S(r_{\alpha\gamma})$$

where

$$U^H(r_{\alpha\gamma}) = \begin{cases} \infty & \text{for } r_{\alpha\gamma} < d_{\alpha\gamma} \\ 0 & \text{for } r_{\alpha\gamma} > d_{\alpha\gamma} \end{cases}$$

TABLE 4-5. Calculated Hard and Soft Ion Pair Friction Coefficients
in 10^{13} kg s^{-1} [a]

System	°C	ζ^H_{+-}	ζ^H_{++}	ζ^H_{--}	ζ^S_{+-}	ζ^S_{++}	ζ^S_{--}
NaCl	800	5.7	0.4	0.5	8.4	3.1	5.1
KCl	800	4.9	1.2	1.1	10.2	2.2	1.4
RbCl	750	8.1	0.0	0.0	14.4	0.7	0.1
CsCl	700	9.0	0.0	0.0	13.9	1.2	0.1
NaI	700	5.3	0.0	0.2	5.9	0.1	37.1

[a]Reprinted with permission from K. Ichikawa and M. Shimoji, *Trans. Faraday Soc.*
66, 843–849 (Table 2) (1970).

For $r_{\alpha\gamma} > d_{\alpha\gamma}$,

$$U^S(r_{\alpha\gamma}) = A_{\alpha\gamma} \exp\left(-\frac{B_{\alpha\gamma} - r}{C_{\alpha\gamma}}\right) - \frac{Y_{\alpha\gamma}}{r^6} - \frac{Q_{\alpha\gamma}}{r^8} + \frac{z_\alpha z_\gamma e^2}{\varepsilon r} \qquad (4.38)$$

ε is the relative permittivity; A, B, C, Y, and Q are constants obtained from
theoretical calculations for ionic crystals. The pair correlation functions
were obtained by resolving the mean distribution function derived from
X-ray diffraction experiments.

The results for ζ^S and ζ^H and the diffusion coefficients of five alkali
halides are shown in Tables 4-5 and 4-6. Of particular interest is the
manner in which ζ^S varies with cation size, indicating that like-ion interactions are very important when the anion is much bigger than the cation.
The agreement between experiment and theory is quite remarkable but
appears to be somewhat fortuitous. If one calculates the diffusion coefficients via the diffusion equation and the friction coefficients in Table 4-5,
then the values obtained do not agree with those in Table 4-6, quoted by

TABLE 4-6. Comparison between Theoretical and
Experimental Diffusion Coefficients in 10^9 m^2 s^{-1} [a]

System	°C	D_+		D_-	
		Theor.	Expt.	Theor.	Expt.
NaCl	800	8.01	8.5	6.02	5.9
KCl	800	6.47	7.2	6.30	6.5
RbCl	750	4.95	4.9	4.95	4.4
CsCl	700	4.52	3.9	4.66	4.4
NaI	700	9.46	7.9	4.02	4.2

[a]Reprinted with permission from K. Ichikawa and M. Shimoji, *Trans. Faraday Soc.*
66, 843–849 (Table 3) (1970).

TABLE 4-7. Calculated Friction Coefficients and
Diffusion Coefficients in Fused KCl[a, b]

T/K	Pressure/kbar	ζ^H_{+-}	ζ^H_{++}	ζ^H_{--}	ζ^S_{+-}	ζ^S_{++}	ζ^S_{--}	D_+	D_-
1043	− 0.9	5.61	0.002	0.05	7.34	0.82	0.99	1.04	1.04
1043	2.4	6.19	0.001	0.03	8.13	0.79	0.93	0.95	0.95
1306	− 0.03	5.06	0.01	0.07	7.58	0.65	0.63	1.35	1.34
1306	1.44	4.95	0.01	0.08	7.86	0.71	0.68	1.35	1.32
1073[c]	0.0	4.9	1.2	1.1	10.2	2.2	1.4	—	—

[a] Reprinted with permission from B. Cleaver, S. I. Smedley, and P. N. Spencer, *J. Chem. Soc. Faraday Trans. I* **68**, 1720–1734 Table 6, (1972).
[b] Values of ζ in 10^{-10} (g s^{-1}). Values of D in 10^{-4} (cm^2 s^{-1})
[c] Values calculated by Ichikawa and Shimoji.

Ichikawa and Shimoji. The quoted diffusion coefficients, however, are still of the right order of magnitude and show the same trends as the correctly calculated figures.

Cleaver et al.[13] have carried out similar calculations for KCl at several temperatures and pressures, Table 4-7. They used the pair potential and distribution function derived from the work of Singer and Woodcock on Monte Carlo and ionic dynamics calculations for fused KCl and other alkali halides. For KCl the friction coefficients are different from those calculated by Ichikawa and Shimoji, and the source of the disagreement is thought to lie with the different pair correlation functions used by the two groups of authors. From the results in Table 4-7 and the preceding equations, Cleaver et al. calculated the diffusion coefficients in KCl and attempted to calculate the electrical conductivity. The diffusion coefficients are displayed in Table 4-8 at several temperatures and pressures. An interesting result of their calculation was that they were able to support an earlier contention of Rice[41] that the long-range coulomb potential should have little effect on the transport properties of a molten salt since the latter are determined only by short-range forces. The integrand of equation (4.35) converges at $r/d_{\alpha\beta} = 1.33 \times 10^{-10}$ so that short-range forces contribute most to the soft friction coefficient. Indeed, if all attractive terms are omitted from the potential, leaving only an exponential repulsive term, the effect is to reduce the calculated diffusion coefficient by only 4%. From Table 4-7 it is clear that $\zeta^H_{\alpha\beta}$ and $\zeta^S_{\alpha\beta}$ make the principal contribution to D_α and D_β, and they are therefore practically equal as are activation energies E_D. Both of these observations concur with experiment, although the magnitude of these quantities do not. The calculated diffusion coefficients are 50–70% too high at 1043 K but only 5–15% too high at 1306 K. This is consistent with the known weakness of equation (4.35), which underestimates negative contributions to the velocity autocorrelation function; this

TABLE 4-8. Calculated and observed Diffusion Coefficients, Activation Energies, and Activation Volumes in Fused KCl[a]

T/K	$10^4 D_+$ (cm^2/s)	$10^4 D_-$ (cm^2/s)	$(E_D)_+$ (kJ/mol)	$(E_D)_-$ (kJ/mol)	$(\Delta V_D)_+$ (cm^3/mol)	$(\Delta V_D)_-$ (cm^3/mol)
1043^b	0.66	0.59	29	30	—	—
1043^c	1.01	1.01	13	13	2.4	2.4
1043^d	0.75	0.75	26	26	—	—
1306^b	1.29	1.18	29	30	—	—
1306^c	1.35	1.34	13	13	1.3	1.3
1273^d	1.30	1.30	26	26	4.6	4.6
1273^d (p = 8 kbar)	0.92	0.92	—	—	± 1.5	± 1.5

[a] Reprinted with permission from B. Cleaver, S. I. Smedley, and P. N. Spnencer, *J. Chem. Soc. Faraday Trans. I*, **68**, 1720–1734 (Table 7) (1972). The values listed are for atmospheric pressure, unless otherwise stated.
[b] Experimental data.
[c] Calculated from data in Table 4-7.
[d] Calculated from computer simulation experiment.

in turn causes the calculated diffusion coefficient to be too high at high densities.

From the published values of Λ, D_α, and D_β, Δ [equation (4.31)] is found to be $+0.22$ for KCl at its melting point.[42, 43] However, the values calculated via equation (4.31) are negative, and $|\Delta|$ is much too large for reasonable values of $d_{\alpha\beta}$, the lower limit of the integral. This result reflects the great sensitivity of the integral in equation (4.31) to the form of the pair potential function. For these calculations the coulombic part of the pair potential was omitted because of the screening effect of oppositely charged neighboring ions. But the main factor determining this integral is the repulsive term of $U(r)$, and the inclusion or omission of attractive terms is of relatively little importance. Thus the error in evaluating Δ could result from small errors in the form of $U(r)$ at small values of r.

Apart from the difficulty of obtaining a reasonable value of Δ, and hence Λ, the theory provides a useful semiquantitative account of ionic mobility in molten salts as a function of temperature and pressure.

At a qualitative level the theory can also account for the variation of ΔV_Λ with ion size, as observed for the alkali halides in Table 4-1.[13] The pressure dependence of Λ arises from the effect of pressure on the friction coefficients and on Δ. The only pressure dependent terms in equations (4.31), (4.35), and (4.37) are the number density ρ and $g^2(r)$. Now the pressure dependence of ρ is proportional to the isothermal compressibility that, for the alkali halides, increases when the size of either ion is increased. $g^2(r)$ is contained within the integral of equations (4.31) and

(4.35), which are determined almost completely in the range of r values for which the curvature of the potential is large. This range includes the first peak in $g^2(r)$. Increasing pressure is expected to increase the area under this peak, both for like and unlike interactions. This ordering effect of pressure will be greatest for the most compressible salts. For lithium salts, in which anion–anion contact probably occurs, the effect of pressure on $g^2_{\alpha\beta}(r)$ will be small, the main result of compression being to increase the area of the first peak of $g^2_{\alpha\alpha}(r)$. For salts of heavier alkali metals, both like and unlike pair distribution functions should be increased by pressure, causing relatively greater increases in $\zeta^S_{\alpha\gamma}$ and Δ. Thus D_α, D_β, and Λ should show greater pressure dependence with increasing ion size, as observed.

On overall balance, the agreement between theory and experiment would suggest that the Rice–Allnatt theory, incorporating their small step diffusion model for the molecular friction coefficient, is a reasonable first-order approximate description of transport in simple liquids and molten salts. However, doubt has been cast on the validity of the fundamental concepts on which the theory is based. The basis of this doubt arises from the analyses of computer simulation experiments of liquid particle dynamics.[25, 44]

According to the theory the relaxation time of the soft forces must be significantly less than those of the hard-core forces, and as a consequence cross correlations between hard and soft forces should be insignificant. Figure 4-5 displays for simulated liquid KCl, the force autocorrelation function resolved into components that correspond to the repulsive F_R, coulombic F_C, and dispersion (noncoulombic attractive forces) F_D forces of the intermolecular pair potential. First, we note that $\langle \mathbf{F}_D(0) \cdot \mathbf{F}_D(t) \rangle$ decays to zero in approximately 5×10^{-14} s, while $\langle \mathbf{F}_R(0) \cdot \mathbf{F}_R(t) \rangle$ passes through a deep minimum at 11×10^{-14} s and also exhibits damped oscillatory behavior. Compared to both of these functions $\langle \mathbf{F}_C(0) \cdot \mathbf{F}_C(t) \rangle$ decays rather slowly to zero in 16×10^{-14} s, the conclusion being that energy and momentum are dissipated by repulsive and dispersion forces before $\langle \mathbf{F}_C(0) \cdot \mathbf{F}_C(t) \rangle$ has changed appreciably. This is a conclusion that supports an earlier contention that coulombic forces play a minor role in determining transport properties of ionic melts. However, it is clear from Fig. 4-5 that while $\langle \mathbf{F}_C(0) \cdot \mathbf{F}_C(t) \rangle$ can be ignored in this analysis, the cross term $\langle \mathbf{F}_R(0) \cdot \mathbf{F}_C(t) \rangle$ cannot since it decays at approximately the same rate as $\langle \mathbf{F}_R(0) \cdot \mathbf{F}_R(t) \rangle$. It is also noticeable that the decay time of the dispersion (soft) forces F_D is less than that of the repulsive (hard) forces F_R, but not significantly less as the theory requires. Consequently force and momentum correlation are carried into successive molecular encounters as evidenced by the negative region in $\langle \mathbf{F}_R(0) \cdot \mathbf{F}_R(t) \rangle$ and the velocity autocorrelation function $\langle \mathbf{v}(0) \cdot \mathbf{v}(t) \rangle$. The negative region in these functions is

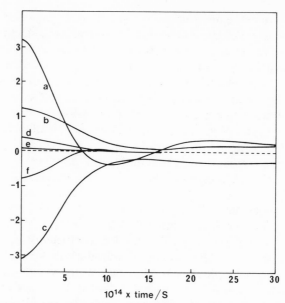

10^{14} x time / S

FIG. 4-5. Resolution of the total force autocorrelation for K^+ ions in liquid KCl at 1045 K. (a) $\langle F_R(0) \cdot F_R(t) \rangle$, (b) $\langle F_C(0) \cdot F_C(t) \rangle$, (c) $\langle F_R(0) \cdot F_C(t) + F_C(0) \cdot F_R(t) \rangle$, (d) $\langle F_C(0) \cdot F_D(t) + F_D(0) \cdot F_C(t) \rangle$, (e) $\langle F_D(0) \cdot F_D(t) \rangle$, (f) $\langle F_R(0) \cdot F_D(t) + F_D(0) \cdot F_R(t) \rangle$. Reprinted with permission from S. I. Smedley and L. V. Woodcock, *J. Chem. Soc. Faraday Trans. II* **69**, 955–966 (Fig. 8) (1973).

thought to arise from a hard-core collision between a molecule and its cage of neighboring molecules that tends to reverse the direction of motion leading to a negative correlation.

A further test of the theory has been to examine the condition that $m/\zeta \gg \tau_1$ for equation (4.27) to be valid. Now τ_1 is the time at which the integrand of the force autocorrelation function reaches a plateau value, sometimes taken to be the minimum in the force autocorrelation function. The friction coefficient for K^+, calculated via equation (4.25) and using the diffusion coefficient (calculated from computer simulation of KCl), has the value of $\zeta = \zeta^H + \zeta^S = 18.4 \times 10^{-10}$ g s^{-1}. Therefore, $m/\zeta \doteq 3.5 \times 10^{-14}$ s, compared with $\tau_1 \doteq 12 \times 10^{-14}$ s, the time of the minimum in the total force autocorrelation function. The condition is not fulfilled. An analysis similar to the above has been carried out on simulated argon,[44] and it confirms the conclusions arising from the study of KCl as to the validity of the Rice–Allnatt kinetic theory. The results indicate that a further step in the development of the theory would be to incorporate cross correlations between hard and soft forces. The "two-event" mechanism for momentum and energy transfer is a poor physical approximation to the dynamics of real liquids.

More recent computer simulation experiments on ionic melts have provided an interesting insight into the origin of the Nernst–Einstein deviation parameter Δ, equation (4.33).[45-47] The electrical conductivity κ can be calculated from the normalized autocorrelation function of the total electric current $J(t)$,[47] defined as

$$J(t) = \left\langle \left[\sum_{i \in \beta} \mathbf{v}_i(t) - \sum_{j \in \alpha} \mathbf{v}_j(t) \right] \left[\sum_{i \in \beta} \mathbf{v}_i(0) - \sum_{j \in \alpha} \mathbf{v}_j(0) \right] \right\rangle \quad (4.39)$$

where α and β refer to anions and cations, respectively, and $\mathbf{v}_i$ is the velocity of ion i and for $1:1$ electrolytes

$$\kappa = \frac{e^2}{3VkT} \int_0^\infty J(t)\, dt \quad (4.40)$$

The diffusion coefficient of α and β ions is calculated from the normalized velocity autocorrelation function $z(t)$, given for α ions as

$$z_\alpha(t) = \langle \mathbf{v}_i(0)\cdot\mathbf{v}_i(t) \rangle \quad (4.41)$$

and

$$D_\alpha = \frac{1}{3} \int_0^\infty z_\alpha(t)\, dt \quad (4.42)$$

with a similar result for β ions.

Using a pair potential similar to that in equation (4.38) Ciccotti et al.[45] have calculated, for six alkali halides, the electrical conductivities and ionic diffusion coefficients, Table 4-9. The agreement between theory

TABLE 4-9. Results of Molecular Dynamics Calculations[a,b]

Salt	V (cm³ mol⁻¹)	T (K)	D_+ (10^{-5} cm² s⁻¹)	D_- (10^{-5} cm² s⁻¹)	κ (s cm⁻¹)	η (cP)	Δ
Lif	15.00	1287	13.6	11.3	12.1	1.14	0.16
					(9.3)		
NaCl	39.10	1262	10.6	9.9	4.2	0.87	0.09
			(14.0)	(10.1)	(4.2)	(0.83)	(0.18)
NaI	57.46	1081	9.4	6.8	2.5	1.08	0.14
			(10.5)	(5.9)	(2.7)	(1.01)	(0.08)
KI	68.97	989	4.5	3.7	1.42	1.06	−0.06
					(1.38)	(1.40)	
RbCl	56.48	1119	5.0	5.2	1.81	0.99	−0.01
			(6.6)	(5.8)	(1.81)	(0.90)	(0.15)
RbI	75.75	1086	4.3	3.5	1.09	1.12	−0.03
					(1.09)	(0.96)	

[a]Reprinted with permission from G. Ciccotti, G. Jacucci, and I. R. McDonald, Phys. Rev. A. 13, 426–436 (Table 2) (1976). Copyright 1976 American Physical Society.
[b]Values in parentheses are experimental results.

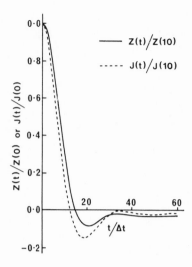

FIG. 4-6. Normalized autocorrelation functions of velocity and electrical current. The dashes show the difference between the two curves. For the system simulating NaCl, $\Delta t = 0.481 \times 10^{-14}$s. Reprinted with permission from J. P. Hansen and I. R. McDonald, *J. Phys. C: Solid State Phys.* 7, L384–6 (Fig. 4) (1974). Copyright The Institute of Physics.

and experiment ranges from 3 to 5%; the values of Δ are in reasonable agreement with the available experimental results for the lighter metal ions but equal to zero for the heavier metal ions. The origin of Δ can be seen from equations (4.33), (4.39), (4.40), (4.41), and (4.42) as arising from cross correlation of velocity in $J(t)$ of the form of $\langle v_i(0) \cdot v_j(t) \rangle, j \neq i$. These contributions show up in the plot of $J(t)$ and $z(t)$ versus t, Fig. 4-6, for simulated NaCl. Comparison of $J(t)$ [equation (4.39)] and $z(t)$ [equation (4.41)] in Fig. 4-6 indicates that the cross correlations are positive, reflecting the fact that motion in the same direction, of a neighboring pair of oppositely charged ions, contributes to diffusion but not to conductance. This does not necessarily indicate the existence of long-lived ion pairs as the correlation need only last for 20×10^{-14} s. A further fact to arise from their analysis is that electrical conductance is not sensitive to the details of the pair potential, such as polarization phenomena and is almost entirely determined by the masses and diameters of the ions. The pair potential used in their work does not allow for ion-induced polarization forces.

4.2.5. The "Free Ion" Theory of Conductance of Barton and Speedy

This theory was developed quite recently and as yet has only been tested against the data for a series of tetraalkylammonium tetrafluoroborate salts.[19] The theory fits the data over a range of temperatures and pressures with only three parameters and appears to predict a glass transition in liquids at high pressures.

The basic postulate is that transport of mass or charge occurs when a fluctuation in local microstructure produces an uncoordinated ion that is then free to translate under the influence of a chemical potential gradient or an external electric field. The self-diffusion coefficient would then be

$$D = D_0 N_0 \qquad (4.43)$$

where N_0 is the number of uncoordinated ions and is a function of temperature and pressure, D_0 is constant. For a square-well fluid of N spheres of diameter d and a square-well potential of depth $-\psi$ and width σ

$$N_0 = N g_0 q_0^t \exp\left(-\frac{\psi}{kT}\right) \Big/ \left[\sum_{z=0}^{c} g_z q_z^t \exp\left(-\frac{z\psi}{kT}\right) \right] \qquad (4.44)$$

The denominator in equation (4.44) is the total partition function of an ion in the liquid, z is the coordination number of the ion, q_z^t the translational partition function, g_z the *a priori* probability of the occurrence of an ion with coordination number z. To estimate g_z it was assumed that there is a linear relationship between the free volume of an ion and its coordination number. Thus, a sphere with no neighbors must have a free volume $v_i > v_f^i$, where $v_f^i \doteq (4/3)\pi[(d/2 + \sigma)^3 - (d/2)^3]$, and if $V_f = \Sigma_i v_i$, $V_f^i = N v_f^i$, then according to Cohen and Turnbull's free volume theory

$$g_0 = \exp\left(-V_f^i/V_f\right) \qquad (4.45)$$

For a sphere with z neighbors, where the close-packed coordination number is c, then

$$V_i \doteq \left(\frac{c-z}{c}\right) V_f^i$$

and

$$g_z = \exp\left(-V_f^i/V_f\right)\left[1 - \exp\left(-V_f^i/cV_f\right)\right]\exp\left(zV_f^i/cV_f\right) \qquad (4.46)$$

By assuming that the Nernst–Einstein equation is valid equations (4.43)–(4.46) give

$$\Lambda = \Lambda_0 T^{-1/2} \Big/ \left\{ 1 + \left[1 - \exp\left(-\frac{V_f^i}{c_f}\right) \right] \frac{x(x^c - 1)}{x - 1} \right\} \qquad (4.47)$$

where Λ_0 is a constant and $x = \exp(V_f^i/cV_f)\exp(E/cRT)$, where $V_f = V - V_s$ and $E = Nc\psi$. Table 4-10 shows the values of the three constants E, V_s,

TABLE 4-10. Parameters in Equation $(4.47)^{a, b}$

Salt	$\ln \Lambda_0$	$(E/R)/K$	$V_s/cm^3 mol^{-1}$	Number	Standard deviation in $\ln \Lambda$
$(n\text{-butyl})_4 NBF_4$	11.1238	2990	288	49	0.018
$(n\text{-pentyl})_4 NBF_4$	11.6752	3740	345	59	0.031
$(n\text{-hexyl})_4 NBF_4$	11.8741	4140	405	53	0.035
$(n\text{-heptyl})_4 NBF_4$	10.7755	3880	463	52	0.029

[a] Reprinted with permission from A. F. M. Barton and R. J. Speedy, *J. Chem. Soc. Faraday Trans. 1*, **70**, 506–527 (Table 7) (1974).
[b] $c = 4$, $V_f = 65 \ cm^3 mol^{-1}$

and Λ_0 that fit the conductance data for four $R_4 NBF_4$ salts. V_s increases linearly with cation volume, as would be expected, being only 3% less than the volume of the solid for the hexyl salt, for example. E/R is an approximately linear function of the 1-atm melting point. The variation in melting point of these salts has been related to the entropy of the alkyl chains of the cations so that E/R might be related in some way to entropy of these chains. This relationship could derive from additional degrees of vibration and rotation that an anion would take up when it becomes uncoordinated from the cation; this would add an extra term to the molecular partition function and hence an extra component to the energy term $Nc\psi$ that would be proportional to the number of extra degrees of freedom. The theory does not attempt to explain the significance of Λ_0, which depends upon the detail of the transport mechanism; however, the authors claim that other empirical equations require at least five parameters to fit the data for one salt so that the fit to equation (4.47) is very good.

According to the square-well model the activation volume for transport is given by

$$\Delta V_\Lambda = \beta V \left(RTV_f^i / V_f^2 - P \right)$$

where β is the isothermal compressibility, V and P are the volume and pressure of the system. When $V_s = V - V_f$ and V_f^i are regarded as constant parameters, this equation fits the data for the ΔV_Λ of $Bu_4 NBF_4$ over a range of temperatures and pressures to within a few percent. The parameters have physically reasonable values, and the equation predicts that ΔV_Λ will decrease with increasing pressure as observed. Presumably at higher pressures, as the glass transition is approached, ΔV_Λ would begin to increase as observed for low-temperature molten salts in Chapter 3; however, there are no data available to test this assumption.

Finally, a more realistic pair potential model has been used to predict how $E_{\Lambda,p}$ and $E_{\Lambda,V}$ would vary with temperature and volume. The potential used was an "L well" model that incorporates a shallow, long-range attractive potential. E_V is predicted to be small and constant at high volumes but to increase rapidly as V decreases beyond a certain value; it does not, however, tend to infinity as predicted by the configurational entropy theory. Instead it is predicted to decrease rapidly from a maximum value as volume continues to decrease. It was suggested that this phenomenon might be regarded as a prediction of the glass transition since a second-order discontinuity must occur if $E_V = 0$ to prevent E_V becoming negative as the volume of the system is decreased.

The theory, using simple but plausible approximations, provides a quantitative interpretation of the temperature and volume dependence of conductance in $R_4 NBF_4$ melts. It is interesting that this simple square well model, which neglects the long-range coulomb potential, can provide a useful interpretation of conductance data in molten salts. The theory is open to further development and should be tested against other systems over a wide range of pressures and temperatures.

4.2.6. Other Theories

The postulates of nonequilibrium thermodynamics have been applied to the study of transport in molten salts.[2, 5] The result is generally a series of phenomenological coefficients that relate to the ionic diffusion coefficients and conductivity of a molten salt, e.g., for salt containing α and β ions[48, 49]

$$\frac{\Lambda}{F} = \frac{L_{\alpha\alpha} + L_{\beta\beta} - L_{\alpha\beta}}{c_i |z_i|}$$

and for a tracer ion γ, of species α,

$$\frac{D_\gamma}{RT} = \frac{L_{\gamma\gamma}}{c_\gamma} = \frac{L_{\alpha\alpha}}{c_\alpha}$$

where D_γ is the diffusion coefficient of the γ tracer ion, and small effects due to isotropic mass differences have been neglected. Unfortunately, this approach has not been very fruitful, at least as far as molten salts are concerned, because of the lack of diffusion data and because of the absence of a theoretical interpretation of the L_{ij}. Laity has developed another formalism that expresses the transport-driving forces in a liquid to the conjugate fluxes via "friction coefficients" γ_{ij}. These are discussed in some detail in reference (5), and the discussion is not reproduced here.

There has been, however, some doubts expressed about the validity of this formalism.[2]

Tomlinson[50] has suggested that the "free ion" model of Rice and Roth[51] might be applicable to molten salts. This model was developed for ionic transport in superionic conductors and proposes that ions are thermally excited from ionic states into conducting energy levels where the ion freely translates for a distance before deexcitation. A preliminary test of the theory against the data for KCl was promising, suggesting an area for further work.

Conclusion

A successful theory of ionic conductance in molten salts has not yet been developed. The kinetic theory of Rice and Berne offers promise but needs modification to allow for cross correlation of forces, and an improved method of calculating the friction coefficients must be developed.

Two more recent theories deserve further investigation—that due to Barton and Speedy and the theory of Rice and Roth. Both share some similarities of approach, i.e., free ion translation, but have a different parametric dependence on temperature and pressure.

Computer simulation experiments are being continually improved to provide an important insight into liquid particle dynamics, and it is anticipated that much remains to be learned from these experiments as they become more sophiscated.

5

Ionic Conductivity in Molecular Liquids and Partially Ionized Molten Salts

5.1. Introduction

Liquids at their 1-atm melting point, that are comprised mostly of uncharged molecules, are regarded as molecular liquids. Their conductivity is low and is believed to arise from the ions produced by the self-ionization of the parent molecules. The ionization constant is generally very sensitive to pressure and temperature, and this is a result of the relatively high compressibilities and expansivities of these liquids compared to ionic melts. The latter arise because of the lack of coulombic cohesive forces. Partially ionized molten salts have conductivities intermediate between those of ionic melts and molecular liquids, reflecting the proportion of dissociated ions to covalent molecules or ion pairs.

5.2. The Temperature and Pressure Dependence of Conductivity

The temperature dependence of the conductivity of a selection of partially ionized molten salts and molecular fluids is displayed in Figs. 5-1 to 5-5. Their conductivities can be compared with those of ionic melts in Fig. 4-1. HgI_2, $HgBr_2$, H_2O, and I_2 [6] can be regarded as molecular fluids because their conductivities are very low at the 1-atm melting point. CdI_2, ZnI, $CuCl$, and $BiCl_3$ are regarded as partially ionized melts. Their conductivities are quite high at the melting point, indicating a substantial degree of ionization, but the temperature and pressure dependence of conductance in these melts is distinctly different to that of fully ionic liquids over a comparable temperature and pressure range. This difference is attributed to the displacement of ionization equilibria with temperature and pressure.

133

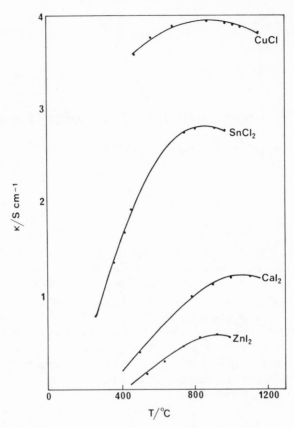

FIG. 5-1. Conductivity of some partially ionized salts versus temperature at constant pressure.[1, 2]

A particular feature of the data in Fig. 5-1 is the negative temperature coefficient of conductivity observed at higher temperatures and, in the cases of HgI_2 (Fig. 5-2) and I_2 (Table 5-2), at the melting point. In accounting for these phenomena Grantham and Yosim[1] argue that the melts contain molecules or ion pairs in equilibrium with ions, the conductivity being the product of the ionic mobility and the degree of dissociation. The former increases with rising temperature, causing the increase in conductivity generally observed at temperatures near the melting point. However, the degree of dissociation decreases with temperature, giving rise to the conductivity maximum and subsequent negative temperature coefficient of conductance. The fall in the degree of dissociation was thought to be principally a consequence of the density change rather than of the temperature change itself. An example is that of $BiCl_3$, at 700°C and at

$\rho = 3.34 \text{ g cm}^{-3}$ the conductivity $\kappa = 0.33 \text{ S cm}^{-1}$, but in the vapor phase at $\rho = 0.16 \text{ g cm}^{-3}$, $\kappa < 10^{-6} \text{ S cm}^{-1}$ at the same temperature. Thus an ionic fluid should tend to a lesser degree of ionization as the density decreases with rising temperature, giving rise to a conductivity maximum. Grantham and Yosim suggested that the test of their hypothesis would be to measure the conductance at constant volume over the same range of temperatures; under these conditions the conductance should always increase.

According to Grantham and Yosim if the activation energy for ionic mobility is approximately the same for all ions, then the temperature of the conductance maximum T_{max} will be determined by the temperature at which the melt becomes gaslike or molecular. They argue that the ease with which a liquid becomes gaslike or molecular should be proportional to the enthalpy of vaporization, and therefore T_{max} should be proportional to $\Delta_f^g H/\text{equiv}$. Table 5-1 displays the T_{max} and enthalpy data for a number of salts, including an estimated T_{max} for NaCl and LI obtained by the extrapolation of conductance data. The relationship is remarkably linear,

TABLE 5-1. Temperature of the Conductivity Maximum and Enthalpy of Vaporization[a]

Salt	$\Delta_f^g H/\text{kcal equiv}^{-1}$	$T_{max}/°C$
NaCl	45.3	$\gg 1075$
LiI	38.7	$\gg 900$
TlCl	24.4	$\gg 900$
TlBr	24.3	$\gg 850$
TlI	25.2	$\gg 1050$
CuCl	6.7	850
$CdCl_2$	15.9	~ 1300
$CdBr_2$	13.8	~ 1250
CdI_2	13.8	1085
$SnCl_2$	9.8	875
$ZnCl_2$	15.4	> 860
ZnI_2	11.5	935
$HgCl_2$	7.0	480
$HgBr_2$	7.1	450
HgI_2	7.1	< 250
$InCl_3$	9.4	< 585
$InBr_3$	8.8	< 436
InI_3	7.8	580
GaI_3	5.4	~ 350
$BiCl_3$	6.2	425
$BiBr_3$	6.2	425
BiI_3	6.6	525

[a]Reprinted with permission from L. F. Grantham and S. J. Yosim, *J. Chem. Phys.* **45**, 1192–1198 (Table 2) (1966).

and the data fit reasonably close to an equation of the form

$$(T_{max} - 273) = (\Delta_f^g H/\text{equiv} - 2)100$$

This relationship may be somewhat fortuitous, as the following analysis will show. Consider the various equilibria postulated for some of the liquids in Figs. 5-1, 5-2, and 5-5:

$$H_2O(aq) \rightleftharpoons H^+(aq) + OH^-(aq)$$

$$2HgI_2 \rightleftharpoons HgI^+ + HgI_3^-$$

$$CdI^+ \rightleftharpoons Cd^{2+} + I^-$$

$$CdI_3^- \rightleftharpoons Cd^{2+} + 3I^-$$

If K is the equilibrium constant for the general reaction $AB \rightarrow A^+ + B^-$, then the conductance κ is given by

$$\kappa = \Lambda°/(a_{AB}K)^{1/2} \tag{5.1}$$

where $\Lambda°$ is the intrinsic molar conductivity of ions in the melt calculated from the ionic mobility and therefore dependent only on temperature and pressure; a_{AB} is the activity of species AB. This gives the equation

$$\left(\frac{\partial \ln \kappa}{\partial T^{-1}}\right)_p = \left(\frac{\partial \ln \Lambda°}{\partial T^{-1}}\right)_p + \frac{1}{2}\left(\frac{\partial \ln K}{\partial T^{-1}}\right)_p$$

where a_{AB} is assumed to remain constant. This will be approximately true for dissociation equilibria that lie well to the left. From the definition of $E_{\Lambda,p}$ and from standard thermodynamics, equation (5.1) can be written as

$$E_{\kappa,p} = E_{\Lambda°,p} + \frac{1}{2}\Delta H° \tag{5.2}$$

where $\Delta H°$ is the standard enthalpy change accompanying the dissociation reaction. Both $E_{\Lambda°,p}$ and $\Delta H°$ are functions of temperature and pressure. For fully ionized melts $E_{\Lambda,p}$ decreases slightly with increasing temperature so that $E_{\Lambda°,p}$ would be expected to behave similarly. The conductance maximum must then occur when $\Delta H°$ becomes negative and equal to $E_{\Lambda°,p}$. For H_2O, $\Delta H°$ is positive at the melting point but becomes negative just before the temperature of the conductance maximum. Thus the maximum is determined by the magnitude and sign of $\Delta H°$. This quantity is quite different to $\Delta_f^g H$, which is the enthalpy change for the processes

(i) $A^+ + B^- \rightarrow AB(l)$
(ii) $AB(l) \rightarrow AB(g)$

Per mole of salt, reaction (i) would contribute very little to $\Delta_f^g H$, most of which must arise from the enthalpy change occurring in reaction (ii). However, for melts that are highly ionized reaction (i) might provide most of the enthalpy change for the vaporization process, i.e., $\Delta_s' H \doteqdot \Delta H°$.

A number of authors have reported high-pressure studies on partially ionized melts and molecular fluids.[3-5, 7-12] They confirm the hypothesis that the drop in conductivity at high temperatures is due to a decrease in density and subsequent decrease in ionic dissociation. The mercuric halides $HgCl_2$, $HgBr_2$, and HgI_2 have been studied by Cleaver and Smedley[7] up to 1 kbar, by Bannard and Treiber[9] to 3.75 kbar and 850°C, by Bardoll and Tödheide[3] to 6 kbar and 500°C, and by Darnell and McCollum[8] to 5.4 kbar for HgI_2 and to 20.5 kbar for $HgCl_2$. The data of Bardoll and Tödheide for HgI_2 (Fig. 5-2) show clearly that at sufficiently high pressures, ~ 3 kbar, the conductivity of HgI_2 increases with increasing temperature, and the increase in dissociation is sufficient to overcome the reduction in ionic mobility even at 6 kbar. Similar results are observed for $HgBr_2$, Fig. 5-3, where increased pressure shifts the conductivity maximum to higher temperatures.

The ultimate conductance experiments, that is the maintenance of constant volume while temperature is increased, have not been performed. Nevertheless, it is possible to calculate $E_{\Lambda,V}$ from high-pressure experiments via the equation

$$E_{\Lambda,p} = E_{\Lambda,V} + (\pi_l + P)\Delta V_\Lambda$$

where $E_{\Lambda,p}$, $E_{\Lambda,V}$, and ΔV_Λ are defined as before [equation (4.9)]. The results from a number of studies are shown in Table 5-2. Note that the activation volumes are invariably negative, indicating that pressure increases the conductivity; for HgI_2 and I_2, $E_{\Lambda,p}$ is negative, but for all the salts $E_{\Lambda,V}$ is positive. The conductivity increases with increasing temperature at constant volume. However, a pressure and temperature must exist where the increase in conductivity due to ionization would be offset by the decrease in ionic mobility at high pressures. Under these conditions ΔV_Λ would be positive, as observed for ionic melts. The trend toward this behavior can be seen in the data of Table 5-2 and from Figs. 5-2 and 5-3. For HgI_2, $HgBr_2$, and $HgCl_2$, ΔV_Λ becomes significantly smaller with rising pressure, and the isobars in Figs. 5-2 and 5-3 become less widely spaced at high pressures. Recent work by Cleaver and Zani[12] show that

TABLE 5-2. Conductivity Parameters, Compressibilities, and Expansivities for Some Molecular Liquids and Partially Ionized Molten Salts[7, 3, 11]

Substance	T (K)	P (bar)	$10^4\alpha$ (K^{-1})	$10^{10}\beta_T$ $(m^2 N^{-1})$	$10^{10}\beta_S$ $(m^2 N^{-1})$	$10^6\Delta V_\kappa$ $(m^3 mol^{-1})$	$10^6\Delta V_\Lambda$ $(m^3 mol^{-1})$	$E_{\Lambda,p}$ $(J mol^{-1})$	E_V $(J mol^{-1})$
AlI$_3$	496	1	7.95	8.9	—	−84	−81	51,830	87,850
	518	1	8.09	8.9	—	−105	−101	51,830	99,400
GaI$_3$	500	1	6.62	7.52	6.63	—	—	—	—
	543	1	6.81	9.17	8.09	−121	−117	13,600	60,780
	566	1	6.92	10.3	9.07	−132	−127	—	—
	583	1	7.00	11.2	9.89	−136	−131	4,800	52,490
InI$_3$	500	1	3.62	6.12	5.47	—	—	—	—
	525	1	3.65	6.60	5.90	−11	−11	14,020	17,240
	574	1	3.72	7.69	6.88	−22	−18	9,910	14,990
	653	1	3.83	10.1	9.02	−28	−23	7,850	13,430
	701	1	3.90	12.1	10.8	−35	−28	6,400	12,700
BiI$_3$	720	1	4.86	5.62	5.03	−6.0	−2.6	8,540	10,160
	754	1	4.94	6.28	5.62	−10.4	−6.5	8,540	12,400
CdI$_2$	661	1	2.54	4.3	—	−0.4	2.0	29,360	28,580
	703	1	2.56	—	—	−1.7	—	—	—
	732	1	2.58	—	—	−3.4	—	—	—
	765	1	2.61	—	—	−3.8	—	—	—
I$_2$	412	1	7.33	5.09	—	−26	−24	−3,940	10,480
	432	1	7.44	5.90	—	−29	−27	−3,940	10,770
	449	1	7.53	6.81	—	−35	−33	−3,940	12,200

Salt	T	P							
HgCl₂	559	1	—	—	—	−45	—	25,700	57,300
	573	1	—	—	—	—	−55	—	68,000
	673	1	—	—	—	—	−88	—	92,000
		2000	—	—	—	—	−48	—	104,000
		4000	—	—	—	—	−39	—	117,000
HgBr₂	514	1	—	—	—	−48	—	25,900	56,100
	588	1	—	—	—	−68	—	19,200	64,400
	723	1	—	—	—	—	—	0.0	73,000
	573	1	—	—	—	—	−64	—	66,000
		1000	—	—	—	—	−52	—	75,000
		2000	—	—	—	−107	−42	—	79,000
	673	2000	—	—	—	—	−58	90,000	100,000
		4000	—	—	—	—	−40	—	100,000
		6000	—	—	—	—	−28	—	90,000
HgI₂	530	1	—	—	—	−90	—	−10,500	35,700
	573	1	—	—	—	−102	—	−13,000	37,700
	623	1	—	—	—	−115	—	−14,700	39,700
	573	1	—	—	—	—	−110	—	45,000
		1000	—	—	—	—	−55	—	28,000
	673	1	—	—	—	—	−156	—	62,000
		2000	—	—	—	—	−37	—	29,000
		4000	—	—	—	—	−11	—	15,000
		6000	—	—	—	—	−5	—	12,000

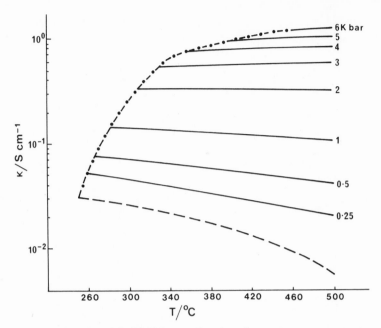

FIG. 5-2. Conductivity of liquid HgI_2 as a function of temperature at constant pressure. $- \cdot - \cdot - \cdot$ solid–liquid coexistence curve; $- - - -$ liquid–gas coexistence curve. Reprinted with permission from B. Bardoll and K. Tödheide, *Ber. Bunsenges Phys. Chem.* **79**, 490–497 (Fig. 3) (1975).

maxima are observed for HgI_2 (at 9.5 kbar, 823 K), CdI_2 (6.5 kbar, 781 K), and $HgCl_2$ (12 kbar, 923 K), and for $BiCl_3$ maxima have been observed at 1 kbar and 300°C, and 3 kbar at 400°C.

In their study of the mercuric halides, Cleaver and Smedley[7] provided further evidence of the effect of pressure on the degree of ionization and refuted any suggestion that the conductance of these melts may be electronic rather than ionic. They carried out electrolysis experiments on HgI_2 to show that at least 80% of the conductance was due to ionic mobility and made a quantitative estimate of the activation volumes ΔV_κ.

Let c denote the concentration, a the activity, and $f_\pm$ the mean ion activity coefficient of the ions HgX^+ and HgX_3^-, where $X = Cl, Br, I$. Then for

$$2HgX_2 \rightleftharpoons HgX^+ + HgX_3^- \qquad (5.3)$$

$$K = a_{HgX^+} a_{HgX_3^-} / a_{HgX_2}^2 \qquad (5.4)$$

and since the ionic concentrations are small in these melts, it is reasonable to assume that a_{HgX_2} will remain approximately constant; therefore,

$$K' = a_{HgX^+} a_{HgX_3^-} = c^2 f_{\pm}^2 \tag{5.5}$$

The standard partial molar volume change ΔV_m°, occurring with ionization (5.1) is given by

$$\Delta V_m^\circ = -RT(\partial \ln K'/\partial p)_T$$
$$= -2RT(\partial \ln c_i/\partial p)_T - 2RT(\partial \ln f_{\pm}/\partial p)_T \tag{5.6}$$

Now if the ionic mobilities are assumed to be independent of pressure and

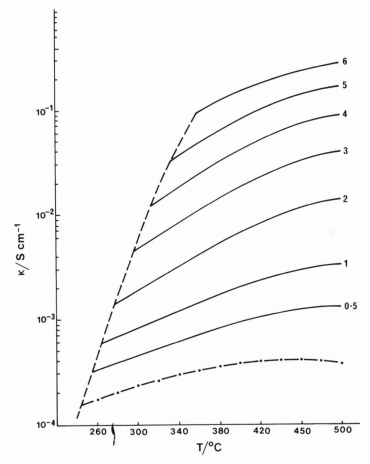

FIG. 5-3. Conductivity of liquid $HgBr_2$ as a function of temperature at constant pressure. $---$ solid–liquid coexistence; $-\cdot-\cdot$ liquid–gas coexistence curve. Reprinted with permission from B. Bardoll and K. Tödheide, *Ber. Bunsenges Phys. Chem.* **79**, 490–497 (Fig. 4) (1975).

the conductance proportional only to the concentration of ions, then

$$-2RT(\partial \ln c_i/\partial p)_T \doteq -2RT(\partial \ln \kappa/\partial p)_T = 2\Delta V_\kappa \qquad (5.7)$$

and from equations (5.6) and (5.7)

$$2\Delta V_\kappa = \Delta V_m^\circ + 2RT(\partial \ln f_\pm/\partial p)_T \qquad (5.8)$$

Thus the hypothesis of increasing ionization at high pressures can be validated if the terms on the right-hand side of equation (5.8) can be evaluated. ΔV_m° is the sum of two terms, the difference between the volumes of the ions and the molecule, and the change in volume due to electrostriction of the solvent when ionization occurs. The former contribution is negligible, $\sim 1 \text{ cm}^3 \text{ mol}^{-1}$; the latter contribution can be calculated from the Born equation for the change in the standard partial molar Gibbs energy ΔG_m°, for solvation of an isolated ion of charge e_i, of radius r_i, in a solvent of permittivity ε,

$$\Delta G_m^\circ = -\frac{N_A e_i^2}{2r_i}\left(1 - \frac{1}{\varepsilon}\right)$$

ΔV_m° is then given by

$$\Delta V_m^\circ = \left(\frac{\partial \Delta G_m^\circ}{\partial p}\right)_T = \sum_i \left[\frac{N_A e_i^2}{2r_i^2}\left(1 - \frac{1}{\varepsilon}\right)\left(\frac{\partial r_i}{\partial p}\right)_T - \frac{N_A e_i^2}{2r_i e^2}\left(\frac{\partial \varepsilon}{\partial p}\right)_T\right] \qquad (5.9)$$

Using various approximations based on the available experimental data, the terms in equation (5.9) have been evaluated and are displayed in Table 5-3; note that the last term is the most significant. The effect of pressure on $f_\pm$ was evaluated from the Debye–Hückel limiting law as

$$RT\left(\frac{\partial \ln f_\pm}{\partial p}\right)_T = \frac{(4.15 \times 10^6)c^{1/2}}{2(\varepsilon T)^{3/2}}\left[-RT\left(\frac{\partial \ln c_i}{\partial p}\right)_T + \frac{3RT}{\varepsilon}\left(\frac{\partial \varepsilon}{\partial p}\right)_T\right]$$

$$(5.10)$$

where the first term in parentheses is given by equation (5.7), and the second term is evaluated from the approximations made in evaluating equation (5.9). On this basis the contribution from equation (5.10) was $-2.4 \text{ cm}^3 \text{ mol}^{-1}$ for $HgCl_2$, $-4.6 \text{ cm}^3 \text{ mol}^{-1}$ for HgBr, and $-50 \text{ cm}^3 \text{ mol}^{-1}$ for HgI_2. The total estimated ΔV_m° are compared with the experimental ΔV_κ values in Table 5-4. Given the assumptions made in the analysis, the agreement is quite good and is slightly improved if the effect of pressure on $f_\pm$ is taken into account. These results substantiate the hypothesis that

TABLE 5-3. Volume Terms For the Species HgX_2, HgX^+, and HgX_3^- [a]

Species	Hg–X bond length $(10^{-10}\ m)$	Molecular volume $(10^{-30}\ m^3)$	Temperature $(°C)$	ε	$10^4(\partial\varepsilon/\partial P)$ (bar^{-1})	$10^6 r^{-1}(\partial r/\partial P)$ (bar^{-1})	$(Ne_i^2/2r^2)\partial r/\partial P(1-1/\varepsilon)$ $(cm^3\ mol^{-1})$	$(Ne_i^2/2r\varepsilon^2)\partial\varepsilon/\partial P$ $(cm^3\ mol^{-1})$
$HgCl_2$	2.29	54.5	286	4.87	4.75	1.6	—	—
$HgCl^+$	2.2	31.2				1.6	6	87
$HgCl_3^-$	2.4	78.7				1.6	1	22
$HgBr_2$	2.41	67.5	241	6.16	7.34	2.0	—	—
$HgBr^+$	2.3	37.6				2.0	7	81
$HgBr_3^-$	2.5	98.2				2.0	2	19
HgI_2	2.59	88.5	257	9.52	20.8	2.3	—	—
HgI^+	2.5	48.0				2.3	8.5	95
HgI_3^-	2.70	129.5				2.3	2	21

[a]Reprinted with permission from B. Cleaver and S. I. Smedley, *Trans. Faraday Soc.* **67**, 1115–1127 (Table 1) (1971).

TABLE 5-4. Volume Terms For the Equilibria $2HgX_2 \rightleftharpoons HgX^+ + HgX_3^-$ [a]

X	Temperature/°C	$\Delta V_m^\circ / cm^3 mol^{-1}$ calculated		$2\Delta V_\kappa / cm^3 mol^{-1}$
		Method 1	Method 2	
Cl	286	− 131	− 116	− 90
Br	241	− 120	− 109	− 96
I	257	− 132	− 127	− 178

[a] Reprinted with permission from B. Cleaver and S. I. Smedley, *Trans. Faraday Soc.* **67**, 1115–1127 (Table 2) (1971).

conductance in these melts is ionic and dependent on the degree of ionization.

Equation (5.8) can be simplified by recognizing that in equation (5.9) the last term in the parentheses makes the biggest contribution to ΔV_m° and by assuming that $f_\pm$ is independent of pressure; the result is

$$\Delta V_\kappa \doteq -N_A \beta_T \sum_i e_i^2 / 12 r_i \qquad (5.11)$$

This equation predicts that the most pressure-sensitive conductances will be observed for those melts with high compressibilities β_T and small ions. It has been further tested against the data for CdI_2, AlI_3, GaI_3, InI_3, and I_2, but satisfactory agreement was obtained for AlI_3, GaI_3, and I_2 only.[11] The failure of equation (5.11) for CdI_2 and InI_3 presumably arises from the neglect of the activity coefficient and from the failure of the Born equation since these compounds have relatively high conductivities and the melts must be extensively ionized at atmospheric pressure.

The experimental data on the conductivity of $BiCl_3$ have been summarized by Treiber and Tödheide[4] (Fig. 5-4). At low temperatures near the melting point the conductivity is quite high, indicating that the melt is substantially ionized. It remains ionized at high pressures even in the supercritical vapor but drops rapidly at low pressures and at supercritical temperatures (905°C). Thus at 1000°C the fluid can be changed from an insulating molecular fluid to an ionic conductor over a range of 4 kbar only.

On the basis that conductance is ionic and arises from the dissociation equilibrium,

$$2BiCl_3 \rightleftharpoons BiCl_2^+ + BiCl_4^-$$

Treiber and Tödheide estimated the degree of dissociation at high temperatures from the ratio of the measured molar conductivity to the value it

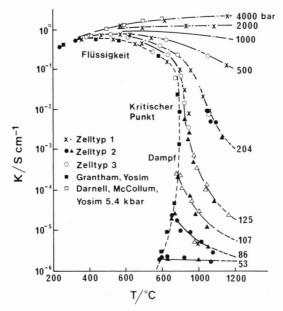

FIG. 5-4.   Conductivity of liquid $BiCl_3$ as a function of temperature at constant pressure. Reprinted with permission from G. von Treiber and K. Tödheide, *Ber. Bunsenges. Phys. Chem.* 77, 541–547 (Fig. 3) (1973).

would have in the absence of association. The latter was calculated by extrapolating the low-temperature high-density conductance data. Table 5-5 displays the equilibrium constant calculated from the degree of dissociation. The very large changes in the equilibrium constant with increasing density substantiate the belief that the most significant contribution to the density dependence of conductivity in these melts arises from the displacement of ionic equilibria.

The conductivity of water has been studied over a greater range of temperatures and pressures than any other liquid and has been extensively reviewed by Tödheide[13] and Holzapfel.[5] Studies range from measurements on water in equilibrium with its vapor to static pressures as high as 100 kbar and 1000°C and shock wave pressures to about 130 kbar and 1000°C, Fig. 5-5. At a given temperature and density the conductivity can be expressed as a product of $\Lambda^{\infty}_{H_2O}$ and k_w,[5]

$$\log \kappa(\rho, T) = \frac{\Lambda^{\infty}(\rho, T)}{10^3} + \frac{1}{2} \log k_w(\rho, T) \qquad (5.12)$$

$\Lambda^{\infty}_{H_2O}(= \lambda^{\infty}_{OH^-} + \lambda^{\infty}_{H^+})$ has been discussed in some detail in Chapter 2; it is

TABLE 5-5. Dissociation Constant for the Reaction $2BiCl_3 \rightleftharpoons BiCl_2^+ + BiCl_4^-$ [a]

T/°C	$\rho/\text{g cm}^{-3}$			
	0.5	1.0	2.0	3.0
800	—	—	—	1.1×10^{-1}
1000	5.3×10^{-10}	8.1×10^{-7}	2.0×10^{-3}	2.0×10^{-1}
1200	1.6×10^{-9}	2.1×10^{-6}	3.1×10^{-3}	3.7×10^{-1}

[a] Reprinted with permission from G. von Treiber and K. Tödheide, *Ber. Bunsenges. Phys. Chem.* **77**, 541–547 (Table 2) (1973).

nearly independent of density and up to 300°C the data fit an empirical formula

$$\frac{\Lambda^{\infty}(T)}{10^3} = 3.4 \times 10^{(-100/T-175)} \tag{5.13}$$

Above 300°C the fit to equation (5.13) is not as good, but is used as an approximate method of extrapolating to higher temperatures. From 0 to 1000°C, Λ^{∞} increases by approximately an order of magnitude, whereas the data in Fig. 5-5 show that κ increases by a factor of 10^8, a consequence of increased dissociation at higher pressures and temperatures. Along the liquid vapor coexistence curve, κ passes through a maximum at ~ 250°C

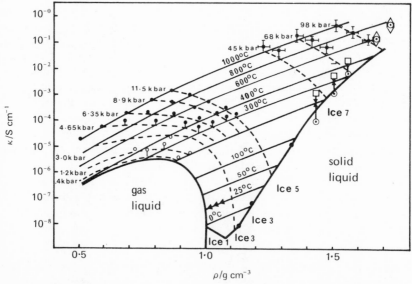

FIG. 5-5. Conductivity of water as a function of density at constant temperature and pressure. Shock wave data $\odot$, $\lozenge$. Reprinted with permission from W. B. Holzapfel, *J. Chem. Phys.* **50**, 4424–4428 (Fig. 6) (1969).

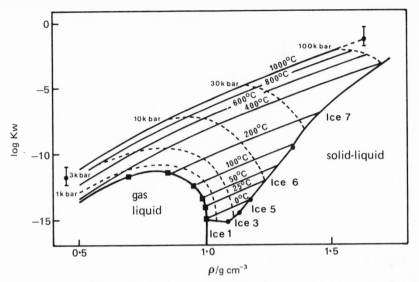

FIG. 5-6. Ionic product of water as a function of density at constant temperature and pressure. Reprinted with permission from W. B. Holzapfel, *J. Chem. Phys.* **50**, 4424–4428 (Fig. 4) (1969).

and in view of the above discussion this must be due to the decrease in k_w shown in Fig. 5-6. Note, however, that initial increase in κ with rising temperature arises from an increase in Λ^∞ and k_w and therefore in the concentration of ions. In fact, through equation (5.12) and (5.13) changes in k_w are mirrored by changes in κ. At the highest pressures the conductivity is high; water is extensively ionized and may be compared to an ionic melt.

A series of empirical equations for k_w and κ allow the calculation of either of these quantities at any temperature and pressure within the range of Figs. 5-5 and 5-6.[5] k_w is given by

$$\log k_w(\rho, T) = 2(7.2 + 2.5\rho/\rho_0)\log(\rho/\rho_0) + \log k_w(\rho_0, T) \quad (5.14)$$

where

$$\log k_w(\rho_0, T) = -(3108/T) - 3.55 \quad (5.15)$$

$$\rho_0 = 1\ \mathrm{g\,cm^{-3}}$$

and therefore $\kappa(\rho, T)$ can be calculated from equations (5.12) to (5.15). In summary, the extensive data on water conductivity and k_w support the hypothesis that self-ionization in molecular fluids increases with increasing pressure and that at constant volume the conductivity always increases with temperature.

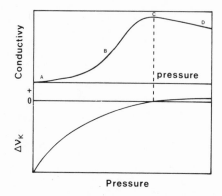

FIG. 5-7.   Schematic diagram showing the variation of the conductivity of a self-ionized
liquid over a wide range of pressure (of the order 0–10 kbar). The corresponding value of ΔV_{κ}
is shown in the lower part of the diagram. The letters A, B, C, and D are referred to in the
text. Reprinted with permission from B. Cleaver, P. N. Spencer, and M. A. Quddus, *J. Chem.
Soc. Faraday I*, **3**, 686–696 (Fig. 8) (1978).

5.3. *Conclusions*

It seems likely that most, if not all, salts exhibit a conductance
maximum if heated to high enough temperatures at low pressures, e.g., the
equilibrium vapor pressure. The temperature may, in some cases, need to
be greater than the critical temperature, e.g., NH_4Cl seems to remain
predominantly ionized when in equilibrium with the vapor to 30° from the
critical point.[14] The maximum is the result of two competing effects: the
increase in ionic mobility and the decrease in ion concentration due to
association at low densities.

The conductance of liquids ranging from molecular to fully ionic have
been studied as a function of pressure. In general, it seems that κ will vary
with pressure according to the curve in Fig. 5-7. Fully molecular melts, for
example $HgCl_2$, fall at point A, partially ionized melts at B (CdI_2), fully
ionized melts at D. The maximum in κ at point C has been referred to
above, and it has been reported that a maximum is observed for HgI_2 at
12 kbar. Thus, it would appear that with the application of sufficiently
high pressure to a molecular liquid, the whole range of conductivities
shown in Fig. 5-7 can be observed. It is likely that many molecular liquids
will undergo extensive ionization at high pressures.

6

Electrical Conductivity in Liquids of Geological and Industrial Interest

6.1. Geological Liquids

The electrical conductivity of geological liquids is usually studied for the purpose of gaining information about the composition of the system or the structure of the liquid. Three types of geological liquids will be referred to in this chapter: the sea, geothermal waters, and magmas or silicate melts.

6.1.1. Seawater

As for all electrolyte solutions the conductivity of seawater is a single-valued function of the pressure, temperature, and ionic strength. The latter property makes the measurement of conductance at known temperature and pressure a valuable method of determining the composition of seawater since the conductance can be measured *in situ*, rapidly, and precisely. The composition or ionic strength of seawater is of great importance to oceanographers because of its effect on the dissolution of minerals, the density, viscosity, adsorption of sound in seawater, and of all of its colligative properties in general.[1]

The relative proportions of the 11 major components of seawater remain very nearly constant, Table 6-1, while the total concentration of solutes can vary considerably. Thus the conductance, which is proportional to the total concentration of ions, can be used to fix the total ionic strength, and from this the concentrations of individual ions may be obtained.

The Ninth Report of the Joint Panel on Oceanographic Tables and Standards[3] has laid down recommended relationships between the composition of seawater (the salinity) and its conductance at a given tempera-

TABLE 6-1. Composition of 1 kg of Natural Seawater of Various Chlorinities[a, b]

Species	$g_i/\text{Cl}(\permil)$	$n_i/\text{Cl}(\permil)$	$e_i/\text{Cl}(\permil)$
Na^+	0.55556	0.0241655	0.0241655
Mg^{2+}	0.06680	0.0027484	0.0054968
Ca^{2+}	0.02125	0.0005302	0.0010604
K^+	0.02060	0.0005268	0.0005268
Sr^{2+}	0.00041	0.0000047	0.0000094
Cl^-	0.99894	0.0281765	0.0281765
$SO_4{}^{2-}$	0.14000	0.0014575	0.0029149
$HCO_3{}^-$	0.00735	0.0001205	0.0001205
Br^-	0.00348	0.0000436	0.0000436
F^-	0.000067	0.0000035	0.0000035
		0.0577772	0.0625179
		$\frac{1}{2}\Sigma = 0.0288886$	$\frac{1}{2}\Sigma = 0.0312590$
$B(OH)_3$	0.00132	0.0000213	0.0000213
$\Sigma = 1.815777$	$\Sigma = 0.0289099$	$\Sigma = 0.0312803$	

With chlorinity C‰, salinity S‰: $S(\permil) = 1.80655(C\permil)^c$

Total mass of sea salt: $g_T = 1.85178\ \text{Cl}(\permil)$

Total moles of sea salt: $n_T = 0.289099\ \text{Cl}(\permil)$

Total equivalents of sea salt: $n_T^1 = 0.0173149\ S(\permil)$

[a] Reprinted with permission from F. J. Millero, in *The Sea*, Vol. 5, Table 1, Wiley-Interscience, New York (1974).
[b] g_i is the number of grams of solute/kg of seawater; n_i is the number of moles of solute/kg of seawater; e_i is the number of equivalents of solute/kg of seawater.
[c] Because of the independent definitions of chlorinity and salinity this relationship is now approximate.

ture at atmospheric pressure. They have recommended the following:

1. The absolute salinity, S_A, be defined as the ratios of mass of dissolved material in seawater to mass of solution. However, in practice this quantity cannot be measured and a practical salinity scale has been devised.

2. The practical salinity scale be based on standard seawater having a conductivity ratio of unity at 15°C to a KCl solution containing a mass of 32.4357 g of KCl in a mass of 1 kg of solution. At this point the standard seawater would have a salinity of 35‰.

3. The practical salinity scale be defined as a function of the conductivity ratios measured at 15°C (R_{15}) of samples prepared from standard seawater of the salinity 35‰ standard after having been diluted by weight

with distilled water and evaporated, according to the following equation:

$$S(\%o) = a_0 + a_1 R_{15}^{1/2} + a_2 R_{15} + a_3 R_{15}^{3/2} + a_4 R_{15}^2 + a_5 R_{15}^{5/2} \qquad (6.1)$$

where the coefficients a are given in the original document and

$$R_{15} = \frac{C(S, 15, 0)}{C(35, 15, 0)}$$

where $C(S, 15, 0)$ is the conductivity of the sample at salinity S, temperature 15°, and zero pressure.

However, equation (6.1) is only useful for unknown samples held at 15°C. At other temperatures they recommend another function (not given here in full)

$$S(\%o) = F(R_T) \qquad (6.2)$$

where

$$R_T = \frac{C(S, T, 0)}{C(35, T, 0)}$$

Thus if the conductivity of standard seawater at the temperature T is known, then it is possible to convert the observed values of the conductivity of the sample to practical salinities. The temperature dependence of the conductivity of standard seawater has been measured and the following relationship obtained

$$r_T = \frac{C(35, T, 0)}{C(35, 15, 0)} = C_0 + C_1 T + C_2 T^2 + C_3 T^3 + C_4 T^4 \qquad (6.3)$$

From equations (6.3) and (6.2) it is possible to calculate the salinity from the conductivity of any seawater sample within the range $1\%o \leqslant S \leqslant 42\%o$ and $-2°C \leqslant T \leqslant 35°C$.

No recommendations were made for the pressure dependence of seawater conductivity since the committee was awaiting more extensive data than that already available.

Thus the conductivity of seawater is known to high precision over a wide range of composition and over a range of temperatures likely to be encountered in the ocean. This knowledge is being extended to include the effect of pressure over that range.

6.1.1.1. The Effects of Concentration, Temperature, and Pressure on Seawater Conductivity

Recent developments in the theory of multicomponent electrolyte solutions have been useful in obtaining an understanding of the thermodynamic properties of seawater.[2] However, the theory of conductance in multicomponent solutions remains to be developed. It was only recently (1977)[4] that a conductance equation for mixed electrolytes was developed beyond the Onsager limiting law. This equation has yet to be tested. Even if this equation proves to be correct in dilute solutions of multicomponent electrolytes, it will not be suitable at the salinity of most seawater, ~35‰, or an ionic strength of 0.7 (Section 3.1). Figure 6-1 shows how the equivalent conductivity of seawater varies with the square root of the equivalent concentration at several temperatures. The equivalent conductivity decreases with increasing concentration but not linearly with $c^{1/2}$ since the concentration range is well beyond that of the Onsager limiting law.

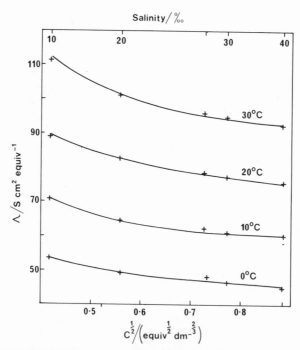

FIG. 6-1. Equivalent conductance of seawater versus $c^{1/2}$. c is the concentration in the total equivalents of salt per dm³.[5, 6]

It would appear from the published literature that no attempt has yet been made to account for seawater conductance in terms of the conductance of its component salts, as given by Table 6-1. However, a start has been made in this direction by Connors and Weyl[7] who have determined the partial equivalent conductivity of sea salt, Λ_S, from the sum of the individual partial equivalent conductivities, Λ_B, determined in seawater. They define Λ_S and Λ_B by

$$\Lambda_B = \lim_{\Delta e_B \to 0} \left(\frac{\Delta G_B}{\Delta e_B} \right) = \left(\frac{\partial G_B}{\partial e_B} \right)_{P,T,n}$$

$$\Lambda_S = \lim_{\Delta e_S \to 0} \left(\frac{\Delta G_S}{\Delta e_S} \right) = \left(\frac{\partial G_S}{\partial e_S} \right)_{P,T} \tag{6.4}$$

where ∂G_B is the increase in conductance when an infinitesimal amount of solution containing ∂e_B equivalents of salt B is added to seawater at a salinity of 35‰, the electrodes of the conductance cell remaining at a constant 1-cm separation. From their experimental data they were able to show that Λ_S given by

$$\Lambda_S = \sum_B \Lambda_B \frac{e_B}{\sum_B e_B} \tag{6.5}$$

was within experimental error of the value of Λ_S calculated from seawater conductivity data. The sum in equation (6.5) is taken over all the salts in Table 6-1 (except for $H_4BO_4^-$) and their respective equivalent fractions. The success of equation (6.5) may result from the fact that the Λ_B are determined by adding the solute B to seawater, so that Λ_B contains terms that arise from the interaction of B with the other solutes. To quote Millero, "it will be interesting in future work to see if ionic conductance data in single electrolyte solutions can be used to predict the conductance of sea salt."

For seawater, the conductivity is given by

$$V\kappa = \frac{\sum\limits_B e_B}{m_{SW}} \Lambda_S + \left[1 - \frac{\sum\limits_B m_B}{m_{SW}} \right] \Lambda_w$$

where m_{SW} is the mass of seawater, m_B the mass of solute B, Λ_w the partial equivalent conductivity of water in seawater, V the specific volume.

The effect of temperature on seawater conductivity at various salinities is displayed in Fig. 6-2. As expected, by comparison with concentrated

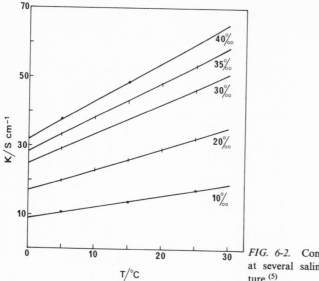

FIG. 6-2. Conductivity of seawater at several salinities versus temperature.[5]

solutions of KOH,[8] KCl,[9] and CaCl$_2$ [9] the conductivity becomes more temperature dependent with increasing salinity. However, again by comparison with other solutions, the equivalent conductivity shows the opposite trend (Fig. 6-1). The increase in conductivity per equivalent of solute due to a temperature rise is not as great in high-salinity water as in low-salinity water. The same phenomenon occurs in aqueous KCl,[8, 10] CaCl$_2$,[9] LiCl, LiBr, and CsCl.[11] At low salinities the increase in conductivity with temperature can be ascribed to the breakdown of water structure, although the increase in conductance will not be directly proportional to the decrease in viscosity (see Chapter 1). A salinity of 35‰ corresponds approximately to 0.5 mol of sea salt per liter of seawater, and at this concentration there can be few water molecules distant by more than three or four molecular diameters from an ion. The structure of the solution and the dynamics of ionic migration are becoming more like those of a low-temperature molten salt (Chapter 3). In fact, seawater (salinity ∼ 35‰) appears to be in the transition region from dilute electrolyte solution to molten salt behavior.

The conductivity of seawater increases with pressure over the measured range of 0–1080 bar; 0–25°C and salinity from 31 to 39‰,[12] Fig. 6-3. Also displayed in this figure is the effect of pressure on a 2.87% by weight NaCl solution, this being the equivalent to NaCl concentration in 35‰ seawater. As expected, the conductivity of seawater increases with increasing pressure, the effect being greater at lower temperatures and at lower salinities. The increase in conductivity of dilute aqueous sodium chloride solutions has been shown to arise almost solely from the positive pressure coefficient of the ionic mobility of the Cl^- ion (Section 2.4.1.1). Thus most of the pressure effect on the conductivity of seawater solutions must arise from the pressure dependence of Cl^- ion mobility. Seawater at a salinity of 35‰ contains more Cl^- ion than a 2.87% NaCl solution, and part of the difference between the two corresponding conductivity curves in Fig. 6-3 may arise from the Cl^- ion composition imbalance. However,

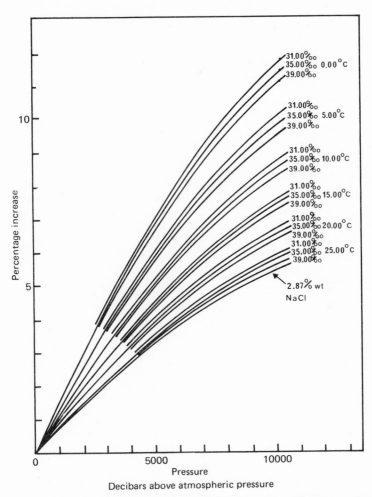

FIG. 6-3. The percentage increase in the conductivity of seawater with pressure. Reprinted with permission from *Deep Sea Research* **12**, A. Bradshaw and K. E. Schleicher, "The effect of pressure on the electrical conductance of sea water." Copyright 1965 Pergamon Press, Ltd.

TABLE 6-2. Molal Dissociation Constants k_m and Standard Partial Molal Volume Changes for Ion Pairs in Water at 25°C As a Function of Pressure[13, 14]

Ion pair	P/bar	k_m	$\Delta V_{B,m}$/cm^3 mol^{-1}
MgSO$_4$	1	0.19	−7.3
	666	0.024	—
	1000	0.026	—
NaSO$_4^-$	1	0.094	−5.4
	1000	0.120	−5.0
	2000	0.149	−4.5
KSO$_4^-$	1	0.096	−8.8
	1000	0.135	−5.9
	2000	0.158	−3.1
MgCl$^+$	1	0.220	−5.9
	1000	0.280	−4.0
	2000	0.320	−2.1

in addition to this must be added the pressure-induced dissociation of ion pairs, which will tend to increase the conductivity beyond that of the comparable NaCl solution. The most significant contributions probably arise from dissociation of MgSO$_4$, NaSO$_4^-$, KSO$_4^-$, MgCl$^+$. Their molal dissociation constants k_m and standard partial molal volume changes on ionization are displayed in Table 6-2; these are all taken from conductance measurements and tend to agree reasonably well with the data from other techniques (Section 2.4.1.1). The total contribution to the conductivity from this effect is very small, however.

From the curves in Fig. 6-3 it is seen that the conductivity decreases with increasing salinity at constant temperature and pressure, i.e., the ratio $\kappa(c)_{P,T}/\kappa(c)_{1,T}$ decreases with increasing concentration c of sea salt. From Fig. 3-2 it can be seen that this behavior is characteristic of a low-temperature molten salt or hydrate melt.

6.1.1.2. Conclusion

The conductance of seawater relative to standard seawater has been precisely measured over a range of salinities, temperatures, and pressures. The qualitative features of the effects of these three variables on seawater conductivity can be interpreted in terms of the equivalent phenomena in binary aqueous solutions. However, it is not yet possible to quantitatively estimate the conductance of seawater from theory, and no attempt appears to have been made to evaluate it from binary solution data.

6.1.2. Geothermal Waters

Geothermal or hot natural waters occur in many parts of the world. Their occurrence, origin, and chemical constitution have been recently reviewed by Ellis and Mahon[15] and, unlike seawater, the composition of geothermal waters can vary considerably. Table 6-3 illustrates the composition of the four main types of geothermal water. These classifications are very broad and serve only as a general classification. Solutions a and b in Table 6-3 are regarded as "alkali chloride waters" because of the high content of NaCl and KCl; solution c is an "acid sulfate water" as the pH is quite low. Hot spring waters containing chloride and sulfate are found in many areas; these are referred to as "acid sulfate–chloride waters," solutions d, e, and f. Solutions high in bicarbonate are known as "bicarbonate waters," solution g. Thus, because of the wide variability in composition, the conductance of geothermal solutions cannot be regarded as an accurate measure of the salinity or salt content of the water, especially if the pH of the solution is unknown, and is of little use to the analytical geochemist. However, the presence of hot saline solutions in porous rocks surrounding a geothermal area can significantly affect the resistivity of the ground. This phenomenon is exploited by the geophysicist in resistivity survey work to outline the area and volume of new geothermal areas. Providing the geothermal water is of the neutral pH alkali chloride type, it is generally assumed to be a dilute sodium chloride solution of a composition that would correspond to its 25°C 1-atm conductivity.[15] From Table 6-3 it can be seen that this would generally correspond to a dilute NaCl solution of < 0.05 m. The conductance of NaCl solutions has been measured over a range of temperatures, pressures, and concentrations, and the interpretation of the data has been given in Chapter 2. No attempts (to the author's knowledge) appear to have been made to measure the conductivity of geothermal waters over a range of temperature or pressure in the laboratory. However, the recent data of Smolyakov[16] on the individual ionic mobilities of some of the ions encountered in geothermal solutions may make it possible to estimate their conductance more precisely. In most cases, their total ionic strength falls within the range of classical electrolyte theory, and by using the techniques described in Chapter 2 for high-temperature solutions it should be possible to obtain a reasonable estimate of the conductance of most of the solutions in Table 6-3 over the range of temperature and pressure encountered in resistivity prospecting, i.e., ~ 80–350°C and up to 300 bar.

Marshall, using the concept of the complete ionization constant (Section 2.3), has discussed how equilibrium constants calculated from conductance measurements can be used to obtain thermodynamic data on

TABLE 6-3. Analyses Typical of Each Water Classification Group[a,b]

Source	pH	Li	Na	K	Rb	Cs	NH$_4$	Ca	Mg	Fe	Mn	F	Cl	SO$_4$	HCO$_3$	B	SiO$_2$	Ref.
a. Geyser 238, El Tatio, Chile	7.32	45	4340	520	6.7	12.6	3.8	272	0.5	0.1	0.4	3.1	7922	30	46	178	260	53
b. Taumatapuhi Geyser, Tokaanu, N.Z.	7.8	22.6	1710	168	0.9	3.3	1.3	32	0.2	0.01	—	1.7	3021	63	2	90	270	53
c. Green-black hot pool, explosion crater, Waiotapu, N.Z.	2.8	—	43	11	—	—	6.2	27	3.5	8.2	—	—	32	347	0	2.5	280	54
d. Yellow hot pool, explosion crater, Waiotapu, N.Z.	2.8	—	405	74	—	—	65.5	40	7.5	5.0	—	—	612	666	0	10.1	370	54
e. Well 205, (1500m deep) Matsao, Taiwan	2.4	26	5490	900	12	9.6	38	1470	131	220	42	7.0	13,400	350	0	106	639	53
f. Crater Lake, Ruapehu, N.Z.	1.20	1.6	740	79	0.04	0.1	11	1200	1030	900	34	260	9450	10,950	0	13.8	852	55
g. Well 5, (471m deep) Wairakei, N.Z.	8.6	1.2	230	17	—	—	0.2	12	1.7	—	—	3.7	2.7	11	680	0.5	191	53

[a]Reprinted with permission from A. J. Ellis and W. A. J. Mahon, *Chemistry and Geothermal Systems* (Table 3.1), Academic Press, New York, N.Y. (1977).
[b]Concentrations in mg/kg (ppm) in waters reaching the surface; pHs measured in cooled waters.

aqueous solutions of geochemical interest.[17] Other studies have been made on aqueous solutions of the dissolved gases usually found in geothermal waters, e.g., SO_2,[18] H_2S,[19] CO_2,[20] and NH_3.[21]

There is a very interesting possibility that a molten saltlike mixture of $NaCl(60-70 \text{ wt}\%)-H_2O$ exists in some geological locations. It is proposed that this fluid could form in equilibrium with phase of 0.05 to 5 wt% NaCl solution at pressures above 500 bar and temperatures above 600°C.[22] Under these conditions the liquid should have a conductivity approaching that of pure NaCl at the melting point. At present, conductance data for very concentrated $NaCl/H_2O$ mixtures under these conditions are not available. However, recent developments in high-pressure, high-temperature techniques should rectify this situation.

6.1.3. Silicate Melts and Magmas

The phenomenon of electrical conductance in silicate melts is of twofold significance. It can be measured precisely, and under conditions of controlled temperature pressure and composition it provides information about the factors controlling the kinetics of mass transport in silicate liquids, which in turn can be used to infer details about liquid structure. Second, the relatively high conductance of molten silicate minerals, in comparison with the solid state, has been suggested as an explanation for the high-conductivity anomalies observed in geomagnetic depth-sounding measurements of the earth's crust and upper mantle.[23] It has been proposed that, under suitable conditions of temperature, pressure, and H_2O content, a region of partially molten material may exist in the crust or mantle, and if the liquid fraction is high enough the conductance of the solid–liquid assemblage will be characterized by the liquid conductivity. With this fact in mind, many studies of the conductance of liquid silicates have been concerned with the large changes occurring on melting of a solid material, rather than the intrinsic factors giving rise to that conductivity.

6.1.3.1. Binary and Ternary Liquid Silicate Melts

In a paper entitled "Electrical Conductivity in Silicate Melts" written in 1948, Bockris and co-workers made the comment, "previous work in the molten state has been mainly technological and *ad hoc* in character and has consisted largely of measurements of complex systems."[24] Apart from their measurements and subsequent papers,[25-27] there appears to be only one other published study of the systematic effects of composition on the conductivity of silicate liquids during the intervening time period.[28]

The conductivity of a range of metal oxide/SiO_2 mixtures has been found to follow an Arrhenius temperature dependence from the liquidus to

TABLE 6-4. Conductivity Parameters for Some Silicate Melts[a]

| Parameter | Dietzel's classification | | | | | | | | | | |
| | Network modifiers | | | | | | | Intermediate | | Network formers | |
Cation	K^+	Na^+	Li^+	Ba^{2+}	Sr^{2+}	Ca^{2+}	Mn^{2+}	Fe^{2+}	Mg^{2+}	Al^{3+}	Ti^{4+}
Cation radius (Å)	1.50	0.95	0.6	1.35	1.13	0.99	0.80	0.75	0.65	0.38	0.35
Coordination number	9	6	6	6	6	6	6	6	6	4	4
Ion–oxygen attraction	0.24	0.36	0.50	0.53	0.63	0.70	0.83	0.87	0.95	1.90	2.61
Composition of melt (M_xO_y/SiO_2)	1:2 1:1	1:2 1:1	1:2 1:1 2:1	1:2 1:1 2:1	1:2 1:1 2:1	1:2 1:1 2:1	1:2 1:1 2:1	1:1	1:2 1:1 2:1	10 wt%	10 wt%
$\kappa_{1750}(S\ cm^{-1})$	1.5 2.4^b	2.1 4.8^b	2.5 5.5 23.2^b	0.18 0.60 1.32^b	0.21 0.63 1.4^b	0.31 0.83 1.15^b	0.55^b 1.8 6.3	1.82^c	0.23^b 0.72 2.15^b	3×10^{-3}	6.3×10^{-4}
A_{1750}	71.8 82.7	83.3 126.0	77.8 109.0 332.0	6.4 16.2 29.9	7.7 15.7 26.8	11.4 18.4 18.8	18.2 35.1 85.5	44.0	6.5 12.2 24.7	0.202	0.051
$\Delta H^{\ddagger}_{1750}$ (kcal/g equiv)	8.2 8.0	12.0 13.5	11.6 10.6 9.6	33.2 17.5 9.0	36.0 26.7 17.0	30.0 20.0 20.0	24.0 16.0 12.0	15.0	34.0 24.0 17.0	22.0	35.7
$\Delta G^{\ddagger}_{1750}$ (kcal/g equiv)	24.6 24.0	24.0 22.3	24.2 22.9 18.4	37.1 33.4 30.8	36.5 33.5 31.4	34.7 32.8 32.7	32.9 30.2 26.6	29.4	37.1 34.5 31.7	52.5	59.3
$\Delta S^{\ddagger}_{1750}$ (cal/g equiv deg)	-8.2 -8.0	-6.0 -4.4	-6.5 -6.1 -4.4	-2.0 -8.0 -10.9	-0.2 -3.4 -7.2	-2.4 -6.4 -11.4	-4.5 -7.1 -7.3	-7.2	-1.6 -5.3 -7.4	-15.3	-11.8

[a] From J. O'M. Bockris, J. A. Kitchener, S. Ignatowicz, and J. W. Tomlinson, Trans. Faraday Soc. 48, 75–91 (Table 3), (1952), with permission.
[b] Extrapolated values.
[c] Extrapolated from Wejnarth's data.

1800°C. Table 6-4 displays the results from these experiments in terms of a Gibbs activation energy $\Delta G^{\ddagger}$, activation enthalpy $\Delta H^{\ddagger}$, and an activation entropy $\Delta S^{\ddagger}$, (Chapter 4). Table 6-5 compares the range and magnitude of the conductivities to those of the equivalent metal chloride. From the fact that the conductivities of silicate melts are similar to those of an ionic melt, that $\Delta G^{\ddagger}$ is positive, and that the faradaic yield on electrolysis is 100% (within experimental error), it is believed that conduction is ionic and not electronic.

Bockris *et al.* interpret their data, shown in Table 6-4, in terms of activation theory and a discrete ion model of liquid silicates. In this model a liquid silicate consists of an assembly of Si^{4+} and O^{2-} ions together with metal cations. Metal ions with a high ion–oxygen attraction $2z_+/(r_{O^{2-}} + r_{M^{2+}})^2$, will tend to bind strongly to oxygen atoms. For example, Si^{4+} coordinates strongly with four O^{2-} ions and forms a chain or network structure if the O/Si ratio is less than four. Silicon is thus a network former; Al^{3+} and Ti^{4+} also fall into this category (Table 6-4). Cations with low ion–oxygen attraction are unable to build a network and are referred to as network modifiers.

Silicate melts containing the monovalent network modifiers Li^+, Na^+, K^+ have similar conductances; in these systems the oxygen ions would be firmly bound to the Si^{4+} ions and the alkali ions would be free to move from one site to the next in the interstices of the silicate network. The enthalpy of activation $\Delta H^{\ddagger}$ is small and unaffected by changes in composition. For the divalent network modifiers, Ba^{2+}, Sr^{2+}, and Ca^{2+} the interpretation of the data is more complex; as the ion–oxygen attraction increases Λ also increases in melts of high SiO_2 composition, but at

TABLE 6.5. Comparison of Conductances of Metal Oxide + Silica Melts and Fused Metal Chlorides[a]

| Metal | Metal oxide + silica melts | | Metal chloride[b] |
	Comp. range (MO/SiO$_2$)	κ_{1750}(S cm^{-1})	κ_{1750}(S cm^{-1})
Li	0.33–0.58	2.35–7.6	8.5
Na	0.19–0.34	0.95–2.8	5.6
K	0.2–0.35	0.66–1.9	4.2
Mg	0.39–0.55	0.33–0.98	2.9
Ca	0.25–0.60	0.25–1.0	5.3
Sr	0.24–0.57	0.09–1.0	4.4
Ba	0.21–0.54	0.04–0.6	4.3
Al	0.018–0.073	0.0006–0.004	0.05 (approx.)

[a] Reprinted with permission from J. O'M. Bockris, J. A. Kitchener, S. Ignatowicz, and J. W. Tomlinson, *Trans. Faraday Soc.* **48**, 75–91 (Table 1) (1952).
[b] Extrapolated to the hypothetical value κ_{1750} according to the formula given by Drossbach.

the orthosilicate composition ($M_xO_y : SiO_2 = 2 : 1$)Λ decreases with increasing ion–oxygen attraction. The latter observations can be explained by the coulombic attraction between M^{2+} and SiO_4^{2-} ions that would become more significant with increasing ion–oxygen attraction and require a large activation energy to separate a cation from one site before it can transfer to the next. This process must also operate in melts above the orthosilicate composition, and the opposite trend with increasing ion–oxygen attraction is attributed to the distortion of the silicate network. The greater the ion–oxygen attraction, the more likely it is to cause a disordering of the rigid silicate network with a consequent increase in cationic mobility. This effect was referred to as "network distortion."

By virtue of their high ion–oxygen attraction the network forming ions Al^{3+} and Ti^{4+} can enter the silicate network that results in their high viscosities and reduced conductances. The activation parameters are high because of the difficulty of activating ions in these melts into the transition state.

The intermediate ions Fe^{2+} and Mg^{2+} display a trend in their properties toward network forming, as evidenced by their relatively high values of $\Delta G^{\ddagger}$ and $\Delta H^{\ddagger}$ compared to other divalent ions.

The entropies of activation are all negative and, according to Bockris *et al.*, this is to be expected on the basis of the model. If $\Delta S^{\ddagger} = (S'' - S')$, where S'' is the entropy of the transition state and S' that of the equilibrium state, then $S' > S''$ if the transition state corresponds to a restricted condition in which the ion pushes through a constriction from one site to the next. For the alkali ions, $\Delta S^{\ddagger}$ is negative, and the network structure will become slightly less rigid on the addition of more metal oxide, i.e., $S'' \rightarrow S'$ because for monovalent cations S' is little changed by network breaking. For the divalent ions, $\Delta S^{\ddagger}$ becomes more negative on the addition of metal oxide. At low ratios of $M_xO_y : SiO_2$, $\Delta S^{\ddagger}$ is small, presumably because when a divalent cation is in an unbroken silicate network it has little freedom, even in its normal state, because it is strongly bound to the network. The decrease in $\Delta S^{\ddagger}$ with addition of metal oxide may arise from a loosening of the structure and an increase in S'.

The applicability of the transition state model to electrical conductance in ionic melts and other ionically conducting systems has been discussed in Section 4.2.1. There is no reason to believe that transition state model should be a valid description of electrical conductance in silicate melts, except possibly at low temperatures approaching the glass transition temperature. Tickle[28] has criticized the application of transition state theory to silicate melt conductance and instead prefers to interpret his and other data in terms of the Cohen and Turnbull free volume theory. From equation (3.7) and the Nernst–Einstein equation (Chapter 1), he obtains

$$\left(\frac{\Lambda T^{1/2}R}{F^2}\right) = \frac{1}{6}\, d\left(\frac{3k}{m}\right)^{1/2} \exp\left(\frac{-\gamma V^*}{V_f}\right) \tag{6.6}$$

where he defines

$$V_f = \frac{V - V_g}{n_i} + V_f^{T_g}$$

V is the specific volume of the melt at T, V_g is the specific volume of the glass at T, n_i is the number of alkali cations per gram, $V_f^{T_g}$ is the average free volume per alkali cation at T_g, the glass transition temperature, and is regarded as an adjustable parameter. A plot of $\log(\Lambda T^{1/2}R/F^2)$ versus $1/V_f$ would have, according to equation (6.6), a slope of γV^*. Tickle argues that the minimum or critical volume γV^* obtained from such a plot should be equal to the volume of an ion V_i obtained from its ionic radius. From a collection of the available data he was able to obtain linear plots of $\log(\Lambda T^{1/2}R/F^2)$ versus $1/V_f$ for Li_2O/SiO_2, Na_2O/SiO_2, K_2O/SiO_2, $Na_2O/K_2O/SiO_2$, BaO/SiO_2, SrO/SiO_2, CaO/SiO_2, and MgO/SiO_2 mixtures. For the Na, K, Ba, Sr, Ca silicates the ratio of $\gamma V^*/v_i$ varied from 1 to 2, in agreement with the free volume theory prediction. At zero V_f, at the intercept of his plots, Tickle was able to obtain the experimental value for the preexponential term in equation (6.6). He found that the ratio of the experimental value to the value calculated from equation (6.6) decreased with decreasing metal oxide content. This was interpreted as being due to a change in the g factor of equation (3.7) from $1/6$ at high concentrations to some lower value; the implication being that, while a cation can jump equally in all three directions at high concentrations, it is restricted to two or one dimension(s) at lower concentrations. However, such restrictions could only arise from an anisotropic melt structure. This seems unlikely in a silicate melt, and his observations are most likely due to the failure of the model. In view of the discussion in Section 3.3 the agreement between the experimental value of γV^* and V_i can only be regarded as coincidental and as having little theoretical foundation.

While neither the free volume theory nor the transition state theory can be regarded as providing an accurate description of electrical conductance in liquid silicates, the transition state formalism provides a better correlation between the properties of the mobile cation (z_i, r_i) and the conductance of the melt.

6.1.3.2. Silicate Melts of Geological Interest

The most systematic study of the electrical conductance of naturally occurring minerals seems to be that of Waff and Weill.[29] They studied the effect of temperature, oxygen fugacity, and composition on the conductance of 10 silicate melts prepared from a range of minerals (Table 6-6). Oxygen fugacity, and hence the ratio of Fe^{2+}/Fe^{3+}, had little effect on the conductance of these liquids, and their conductances all increased with increasing temperature indicating that the conductance was principally ionic. The plots of $\log \Lambda$ versus $1/T$ obtained from their measurements are linear for the HA, HA($2N$), HA($3N$) compositions but show distinct curvature for all the other compositions. Figure 6-4 displays the activation energies as a function of temperature for representative samples. Waff and Weill note that non-Arrhenius temperature dependence occurs for those melts with a high M^{2+}/M^+ ion ratio, and Arrhenius dependence occurs for ions with a low M^{2+}/M^+ ratio. In the former case they propose that the melt conductance will be determined mostly by divalent ions, and because of their high surface charge density and ability to polarize surrounding oxygen atoms, the mobility of these ions will be strongly dependent on the distribution of polymeric silicate anion sizes and hence on temperature. On the other hand, univalent ion mobility should be less dependent on the silicate anion structure and therefore of temperature. An alternative, but probably equivalent explanation, would be to note that many liquids display non-Arrhenius behavior below a temperature of $\sim 2T_g$ (Section 3.3.3). Binary silicate melts of alkaline earth metal oxides have glass transition temperatures of ~ 700–$800°C$, whereas binary silicate melts of alkali metal oxides have glass transition temperatures of ~ 400–$500°C$.[28b] Therefore melts that show a non-Arrhenius temperature dependence are approaching T_g, and their temperature dependence is best explained in terms of a theory that is based on a cooperative mechanism, e.g., the configurational entropy theory. The melts that are low in M^{2+} are probably well above T_g at the temperatures at which this study was conducted and would be expected to show an Arrhenius temperature dependence.

The effect of pressure on the conductance of silicate melts appears to have been studied as an adjunct to the study of melting phenomena.[30] Figure 6-5 shows the effect of pressure on solid, partially molten, and liquid basalt. According to this work the effect of pressure on molten basalt is to cause an increase in the conductivity. This result contradicts the findings of Khitarov and Slutsky,[31] but it is in accordance with the findings of Kushiro *et al.*, who observe that the viscosity of albite $(NaAlSi_3O_8)$,[32] jadeite $(NaAlSi_2O_6)$,[33] basalt, and andesite[34] decrease with increasing pressure. This phenomenon was attributed in part to the

TABLE 6-6. Electron Probe Analyses (wt%) of Glasses Quenched from Post-Run Melts of Tholeiite Basalt (70-15, PG16), Alkali Olivine Basalt (BCR2), Latite (V-31), and Andesite (HA)[a,b]

Oxide	70-15	PG16	BCR2	V-31	HA	HA(2N)	HA(3N)	HA(3F)	HA(6F)	HA(10F)
SiO_2	47.1	47.5	53.9	61.8	57.9	57.6	57.0	57.1	55.4	53.4
TiO_2	3.03	1.54	2.58	0.79	0.63	0.63	0.62	0.62	0.60	0.58
Al_2O_3	15.2	15.1	13.3	14.2	19.0	18.9	18.6	18.7	18.2	17.5
FeO	12.2	14.6	13.3	9.1	4.99	4.96	4.90	7.83	10.4	13.7
MnO	0.21	0.13	0.02	0.15	0.09	0.09	0.09	0.09	0.09	0.08
MgO	7.58	6.91	3.57	0.38	3.10	3.08	3.05	3.05	2.97	2.86
CaO	9.5	9.8	7.5	3.50	7.7	7.6	7.5	7.5	7.33	7.06
Na_2O	2.59	2.95	3.44	4.62	4.15	6.18	7.18	4.09	3.97	3.83
K_2O	0.74	0.60	1.74	4.46	1.06	1.05	1.04	1.04	1.01	0.98

[a] Reprinted from H. S. Waff and D. F. Weill, *Earth Planet Sci. Lett.* **28**, 254–260 (Table 1) (1975) with permission from Elsevier Scientific Publishing Co.

[b] Compositions HA(2N), HA(3N), HA(3F), HA(6F), and HA(10F) correspond to base composition HA with additions of 2.1% Na_2O, 3.2% Na_2O, 3.0% FeO, 6.0% FeO and 10% FeO, respectively. Total Fe is arbitrarily given as FeO.[a]

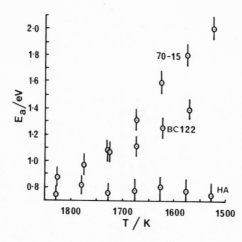

FIG. 6-4. Activation energy as a function of temperature for some silicate melts. The composition of these melts are given in Table 6-6. Data taken from ref. 29.

change in coordination of Al^{3+} from four at low pressures, where it would be incorporated in the silicate network, to six at high pressures, which would require chain rupture. The viscosity of the melt would decrease accordingly. Similar arguments were advanced for Si^{4+} in a $Na_2Si_3O_7$ melt, which also undergoes a small decrease in viscosity at high pressures.[33] Any breakdown in melt structure is likely to lead to enhanced ionic mobility and an increase in conductance. A more systematic study of the

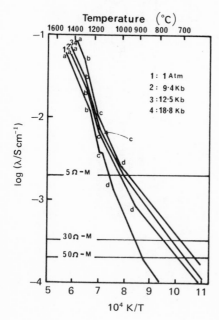

FIG. 6-5. Effects of pressure and melting on the electrical conductivity of dry basalt. Sections b–c of the curves are melting intervals. Reprinted with permission from H. Watanabe, *Spec. Contr. Geophys. Inst. Kyoto Univ.* **10**, 159–170 (Fig. 6) (1970).

TABLE 6-7. Electrical Conductivity as a Function of Water Content for Water-Saturated Silicate Melts under Pressure (κ/S cm^{-1}; percent H_2O by wt)[a]

Material	$T/°C$	1 atm "dry"	Pressure of water/atm				
			1000	2500	4000	7000	10,000
Granite	1000	1.5×10^{-5}	2.8×10^{-2}	1.0×10^{-1}	1.4×10^{-1}	1.6×10^{-1}	1.8×10^{-1}
			3.6	5.8	9.9	14	
	1200	1.3×10^{-2}	1.0×10^{-1}	1.5×10^{-1}	1.9×10^{-1}	2.8×10^{-1}	2.8×10^{-1}
			3.3	5.4	10.4	14.5	18
Basalt	100	4×10^{-5}	5.8×10^{-3}	4.3×10^{-2}	8×10^{-2}	1.5×10^{-1}	1.6×10^{-1}
			2.1		6.0	10.5	13.5
	1200	4×10^{-3}	6.4×10^{-2}	2.4×10^{-1}	2.6×10^{-1}	3×10^{-1}	
			3.3	5.1	7.0	11.5	
SiO_2	1200					1×10^{-3}	4.1×10^{-3}
						7.3	23

[a] Reprinted with permission from E. B. Lebedev and N. I. Khitarov, *High Temperature High Pressure Electrochemistry in Aqueous Solutions*, N.A.C.E. at University of Surrey, England (1973).

effect of pressure on silicate melt conductance may help to improve our understanding of the structural breakdown of silicate melts at high pressure.

Water has a significant effect on silicate melt conductivity (Table 6-7).[35] Up to 4000 atm of water pressure the conductivity increases rapidly, but it levels out above this pressure. It is believed that the water reacts with the oxygen in the silicate chains, rupturing the network with a resultant decrease in viscosity. After the initial breakdown in structure resulting from the addition of ~4 wt% of water, the cation mobility is enhanced to the point where further breakdown of the silicate network has little effect.

6.2. Industrial Electrolytes

There exists a very wide range of industrial electrolytes, and the complete description of them is well beyond the scope of this book. The interested reader is referred to the comprehensive review entitled "Industrial Electrochemical Processes" for further information on the types of electrolytes suitable for industrial use.[36]

6.2.1. Aqueous Industrial Electrolytes

The concentration of most, if not all, industrial aqueous electrolyte solutions is beyond the maximum concentration at which classical electrolyte theory is valid, an example being the production of hydrogen and oxygen from the electrolysis of concentrated KOH solutions.[37] The electrolysis cell is usually operated at 80–85°C with an electrolyte of 25–30 wt% KOH($\sim$5.4 molal). Figure 6-6 displays the conductivity of NaOH and KOH solutions as a function of concentration, and it is clear that the concentration of the industrial electrolyte is chosen to coincide with the conductivity maximum. Maxima, as a function of concentration, have been observed in other electrolyte solutions, e.g., $LiNO_3/H_2O$. Angell proposes that this conductivity maximum can be explained in terms of the configurational entropy theory of mass transport.[38] Equation (3.23) can be written in terms of conductivity κ as

$$\kappa = 1000 \, c \, A T^{-1/2} \exp\left(- \frac{B}{\left[T - \left(T_0^1 + bP + Yx \right) \right]} \right) \qquad (6.7)$$

At constant temperature and pressure and at low values of concentration and therefore mole fraction x, the exponential term in equation (6.7) will

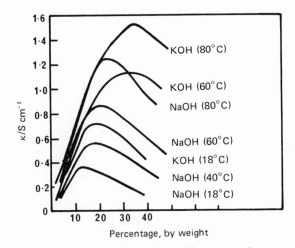

FIG. 6-6. Variation of conductivity of electrolyte with concentration at constant temperature. Reprinted with permission from C. E. Bowen, *Proc. Institution of Electrical Engineers* **90**, 474–485 (Fig. 1) (1943).

be constant, κ will increase as c the concentration of the salt increases. The composition dependence of the exponential term will become significant only in the regime where T_0 becomes composition dependent, Fig. 3-4. Above this composition the exponential will decrease with increasing concentration as shown by equation (6.7). The conductance will pass through a maximum and continue to drop with increasing concentration as the solution approaches T_0. The maximum represents the transition to molten saltlike behavior.

Lown and Thirsk,[39] in their study of the conductance of concentrated LiOH, NaOH, KOH solutions, claimed that the maximum in Fig. 6-6 was due to two factors: (i) ion pair formation as the coulombic energy of attraction increases at high concentrations; and (ii) the gradual disappearance of the special defect proton conductance mechanism (Fig. 3-25). The latter is certainly true, and increasing coulombic energy of attraction between ions is probably one of the factors that gives rise to an increase in T_0.

Chlorine is produced electrolytically from a saturated NaCl solution at $\sim 80°C$ and sodium chlorate from a solution that is ~ 4-molal NaCl and ~ 4-molal $NaClO_3$. Both of these solutions are beyond the transition to molten saltlike behavior and should exhibit the conductance properties of a low-temperature molten salt (Chapter 3). Detailed studies of their conductance properties are not available.

Electrowinning of metals, such as Cu, Co, Zn, Cd, and Mn, is carried out from relatively dilute sulfate solutions.[36] These solutions are generally formed by leaching the metal from the ore with H_2SO_4 solutions, and, after subsequent treatment, this results in an approximately 1–2 molal solution of the metal sulfate, sometimes containing a similar concentration of sulfuric acid. Electrolysis is carried out at approximately 30–60°C, depending on the process. Again a detailed study of the conductance of concentrated divalent metal sulfates at high temperatures is unavailable. However, it is expected that the conductance of these solutions will pass through a maximum that will occur at lower concentrations than that for 1 : 1 electrolytes. The shift of the conductance maximum to lower concentrations with increasing ionic charge would be in accordance with the above interpretation of the phenomenon and the discussion in Chapter 3.

Aqueous industrial electrolyte solutions are generally designed to assist in maximum rates of charge transfer across the electrolytic cell. The composition of many of them corresponds closely to the concentration of the conductance maximum, which in turn represents the transition to low-temperature molten-saltlike behavior where the conductance is controlled principally by the composition dependence of T_0.

6.2.2. Nonaqueous Electrolyte Solutions

The solubility of salts, and the resultant conductivity of ions in organic solvents is low, consequently few organic electrolytes are used in industrial processes. However, a system that has received recent renewed attention is that of KBr or LiBr and Al_2Br_6 in toluene.[40] This system possesses a fairly high conductivity, ~ 0.006 S cm^{-1} in concentrated solutions, and can serve as an electrolyte for the electroplating of aluminum. The molar conductivity as a function of the molar concentration of KBr and LiBr is shown in Fig. 6-7. In very dilute solutions the molar conductivity decreases with increasing concentration, as predicted by classical theory, the principal charge carrying species being presumably M^+ and Br^-. As the concentration increases, first ion pairs and then ion triplets will form, the latter giving rise to the minimum and subsequent rise in conductivity, Section 2.4.2. However, the rapid rise in conductivity, beyond about 7×10^{-3} mol dm^{-3} of MBr, is thought to arise from a completely different process. Reger *et al.*[41] have shown that in this concentration region the principal ions are formed as

$$3MBr + 3Al_2Br_6 \rightleftharpoons \left[M_2(Al_2Br_7) \right]^+ + \left[M(Al_2Br_7)_2 \right]^- \qquad (6.8)$$

the equilibrium lying well to the right because alkali metal bromides are

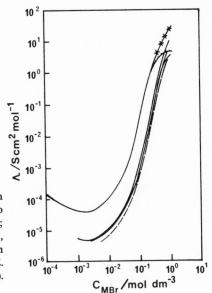

FIG. 6-7. The molar conductivity as a function of concentration for different ratios to Al_2Br_2 to MBr. From top to bottom: Al_2Br_8 : KBr = 1.0; Al_2Br_6 : LiBr = 1.0; Al_2Br_6 : LiBr = 0.75. ×, Curves corrected for viscosity. Reprinted with permission from A. Reger, E. Peled, and E. Gileadi, *J. Phys. Chem.* **83**, 869–873 (1979). Copyright 1979 American Chemical Society.

very insoluble in toluene. The Stokes radii of these ions calculated from the molar conductivity are much smaller than would be anticipated on the basis of their structure, implying that a hydrodynamic mechanism is not the only mode of charge transport in these solutions. Reger *et al.* propose a chain charge-transfer mechanism of the following kind

$$A^+ + B^- (A^+ B^-)_{n-1} A^+ \rightarrow A^+ \text{---} B^- (A^+ B^-)_{n-1} \text{---} A^+$$
$$\rightarrow A^+ B^- (A^+ B^-)_{n-1} + A^+$$

where A^+ and B^- refer to the ions formed in equation (6.8), and $(A^+ B^-)_n$ to neutral aggregates or clusters of ions. At the highest concentrations studied, the molar conductivity seems to be heading for a maximum, or at the least to be leveling off very rapidly. This was attributed to the rapid decline in hydrodynamic mobility of the ions, which controls the rate of charge transport between clusters as the solution becomes more viscous.

6.2.3. Molten Salt Electrolytes

Industrial molten salt electrolytes are generally mixtures of two or three major components.[36] They are used principally for electrowinning and electrorefining of metals, especially those metals that react readily with oxygen or water, e.g., Li, Na, and Al. Their conductivities are high, and they are able to dissolve large amounts of reactant material.

A large body of data exists on the conductivity of molten salt mixtures, but in theoretical terms their conductance properties are not well understood. This is hardly surprising in view of the fact that a suitable theory does not yet exist for conductance in one-component melts. However, some progress has been made in the qualitative interpretation of the deviations from additivity that occur in mixtures of salts. The reader is referred to the discussion by Moynihan[42] for more detail on transport in molten salt mixtures, and the discussion in this book will be limited to the case of some industrial molten salt mixtures.

Lithium is produced by electrolysis from a molten LiCl(60 mol%)/ KCl(40 mol%) eutectic, over a range of temperatures from 400 to 550°C. The equivalent conductivity of LiCl/KCl mixtures are displayed in Fig. 6-8. A feature characteristic of many molten salt mixtures is the negative deviation from additivity of the conductivity isotherms, i.e., the conductivity is less than the sum of the equivalent conductivities multiplied by the equivalent fraction.[44] In systems where complex ion formation is negligible or nonexistent, the negative deviations are thought to arise from a polarization effect. In a mixture of LiCl/KCl some Cl^- ions will have Li^+ and K^+ as immediate neighbors; the smaller Li^+ will exert on the Cl^- an electric field of greater intensity than will a K^+. The Cl^- will become polarized in the direction of the Li^+, strengthening the attractive force between the two ions at the expense of electrostatic attraction between the K^+ and Cl^-. The assumption is that the increased degree of attraction between the Li^+ and Cl^- ions will result in a drop in Li^+ internal mobility with respect to the Cl^-, while the converse should be true for the K^+ ion. Thus the addition of LiCl to KCl should increase the K^+ internal mobility with respect to the Cl^-, and conversely for Li^+. This prediction concurs with experimental measurements of Li^+ and K^+ mobilities in nitrate and chloride mixtures.[42] The minimum in the equivalent conductivity curve in Fig. 6-8 probably arises from the large differences in the composition dependences of the Li^+ and K^+ internal mobilities (with respect to Cl^-) since neither ion mobility passes through a minimum.[43]

Moynihan has summarized the data for a variety of binary mixtures to produce two important generalizations about monovalent cation internal mobility in mixtures containing a common anion:

1. An increase in the mole fraction of the salt with the larger cation will result in a decrease in both cationic mobilities.

2. The rate at which the mobility decreases with addition of the larger cation will be greatest for the smaller cation.

Magnesium has been produced electrolytically from several types of molten salt electrolytes.[36] The most common electrolyte is a mixture of

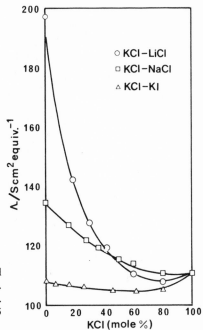

FIG. 6-8. Equivalent conductivity of LiCl/KCl mixtures. Reprinted with permission from E. R. van Artsdalen and I. S. Yaffe, *J. Phys. Chem.* **59**, 118–127 (Fig. 6) (1955). Copyright 1955 American Chemical Society.

varying amounts of $MgCl_2$, $CaCl_2$, NaCl, and KCl plus other minor components. Pure $MgCl_2$ melts at 708°C, and its conductivity is not very high, $\sim 30\ S\,cm^{-2}\,equiv^{-1}$; the addition of $CaCl_2$, NaCl, and KCl lowers the solidus temperature, while NaCl and KCl help to provide a high conductivity. $MgCl_2$/KCl mixtures have been studied over the complete range of composition, and deviations from additivity in this melt have been attributed mostly to the formation of $MgCl_4^{2-}$ ions in equilibrium with Mg^{2+} and Cl^- ions.[45] The data displayed in Fig. 6-9 substantiate the above hypothesis. Mg has a low mobility with respect to the Cl^- ion, which decreases on the addition of more Cl^- as $MgCl_4^{2-}$ ions are formed. Presumably complex ions are present in $MgCl_2$/KCl/NaCl mixtures. $CaCl_2$/NaCl and CaCl/KCl mixtures display similar deviations from additivity, and this has been attributed to complex ion formation.[46] In $MgCl_2$/CaCl/NaCl/KCl mixtures both the Mg^{2+} and Ca^{2+} would compete for Cl^- ions; this may increase the free Mg^{2+} concentration above that which it would have in the absence of Ca^{2+}.

Alkali and alkaline earth metal chlorides are also used as electrolytes for the electrowinning of Na, Be, Ti, B. Na can be produced from a bath of Na, Ca, Ba, and Sr chlorides[47] or Li, K, and Na chlorides,[48] although it is generally won from a Na, Ca, Ba chloride mix.[36] Be may be produced

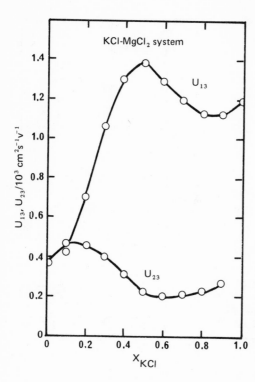

FIG. 6-9. Internal mobilities of K⁺ (U_{13}) and Mg²⁺ (U_{23}) ions with respect to Cl⁻ ions in KCl/MgCl₂ mixtures. Reprinted with permission from W. K. Behl and J. J. Egan, *J. Phys. Chem.* **71**, 1764–1769 (Fig. 4) (1967). Copyright 1967 American Chemical Society.

from $BeCl_2$, although the established method seems to be from a KCl/NaCl melt containing about 14 wt% of $BeCl_2$.[36] Conductivity measurements in the $BeCl_2$/KCl system have been interpreted in terms of complex ion formation, e.g., ions of the type $BeCl_4^{2-}$, $BeCl^+$.[49]

The electrolytic production of aluminum from cryolite melts must be one of if not the most extensively studied electrochemical processes. In their excellent text *Aluminium Electrolysis*[50] Grjotheim *et al.* discuss the physical chemistry of aluminum production, including a section on the electrical conductivity of electrolytes used in the process. The reader is referred to that text for a more complete description of the electrolyte; the discussion in this book will be concerned more with the interpretation (limited as it is) of some of the observed conductance phenomena.

Beside the basic electrolyte Na_3AlF_6/AlF_3/Al_2O_3, substances are added to improve its physico chemical and electrochemical properties, e.g., MgF_2, which assists in separation of carbon particles from the melt, and LiF to enhance its conductivity. The equivalent conductivity of some of the salts that comprise the industrial electrolyte are shown in Fig. 6-10. Potassium and lithium fluorides have a distinctly higher conductivity than

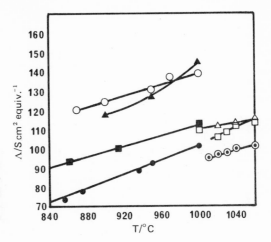

FIG. 6-10. Equivalent conductivity of Li, Na, and k fluorides and cryolites.[51] ○ LiF; ▲ KF; ■ Li_3AlF_6; ● 60% Li_3AlF_6 + 40% Na_3AlF_6 (by Wt.); △ NaF; □ K_3AlF_6; ⊙ Na_3AlF_6. Reprinted by permission of the publisher, The Electrochemical Society, Inc.

sodium fluoride, but even these are less than those of the corresponding chloride melts. The reason for this is not known. However, the cryolites fall into the anticipated order. If the melts consist predominantly of M^+ and AlF_6^{3-} ions, then the much smaller cation should carry most of the current with the smallest ion being able to move with most ease through the almost stationary anion quasisublattice. However, AlF_6^{3-} is thought to undergo ionization, $AlF_6^{3-} \rightleftharpoons AlF_4^- + 2F^-$ where the degree of dissociation at 1000°C is ~ 0.3.[50] Therefore, F^- must also contribute to the total conductance, but the extent to which it does so is unknown.

Addition of metal halides to sodium cryolite can have a significant effect on the conductivity, Fig. 6-11. The observed trends are in the direction that would be predicted on the basis of the approximate additivity of the conductivities. Grjotheim *et al* assume that the increase in conductivity on the addition of NaCl to cryolite is due to the increasing concentration of Na^+ in the melt. The addition of 25 wt% (55 mol%) NaCl to Na_3AlF_6, however, results in a 10% decrease in density at 1000°C and an increase in the volume per mole of Na^+, i.e., a decrease in Na^+ in concentration. The increase in conductivity must arise from some other effect, such as the disruption of the AlF_6^{3-} quasisublattice by Cl^- ions and a subsequent increase in ionic mobility and/or an increase in the number of F^-. It has been claimed that the addition of NaCl to cryolite melts results in an increased degree of dissociation of AlF_6^{3-}.[50] Both LiF and NaCl reduce the viscosity of sodium cryolite, but the addition of CaF_2 increases the viscosity. These results are in keeping with the changes in conductivity.

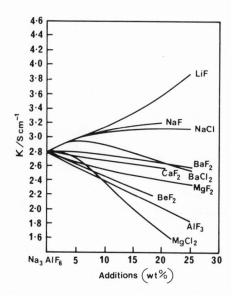

FIG. 6-11. Conductivity of molten Na_3AlF_6/MA_x mixtures at 1000°C. Reprinted with permission from K. Grjotheim, M. Malinovsky, and K. Matiasovsky, *J. Metals* **21**, 28–33 (Fig. 2) (1969). Copyright 1969 American Institute of Mining, Metallurgical and Petroleum Engineers, New York.

The molar conductivity of cryolite drops sharply on the addition of Al_2O_3, e.g., at 1010°C pure Na_3AlF_6 has a conductivity of 284 S cm^2 mol^{-1}; 4 mol% Al_2O_3/94 mol% Na_3AlF_6, 260 S cm^2 mol^{-1}; 15 mol% Al_2O_3/ 85 mol%, Na_3AlF_6 217 S cm^2 mol^{-1}.[52] It has been postulated that Al_2O_3 reacts with cryolite to form a mixture of Na^+, AlF_y^{3-y}, $AlOF_x^{1-x}$ ions where $3 \leqslant x \leqslant 5$ and $4 \leqslant y \leqslant 6$. Presumably the drop in conductivity is caused by the reduction in the number of free F^- ions, a drop in ionic mobility, in part as a result of the increased viscosity and the dilution of charge concentration on the addition of Al_2O_3.

Addition of NaCl to Na_3AlF_6/Al_2O_3 melts results in an increase in conductivity, although the increase becomes less pronounced at high Al_2O_3 content. However, the ternary system $Na_3AlF_6/NaCl/AlF_3$ does not behave in such a regular fashion. In Na_3AlF_6 melts of low AlF_3 content ($\sim$5 wt%) the conductivity increases with NaCl content, but above 5 wt% $AlCl_3$ the conductivity decreases with increasing NaCl content. An explanation for this phenomenon is not available.

It is clear that while many conductance measurements have been made of cryolite melts and the various other components of the industrial electrolyte, the interpretation of the data is in a very underdeveloped state. To develop only a qualitative understanding will require knowledge of the effects of temperature and composition on ionic equilibria, the distribution of ions in the melt, and the individual ionic mobilities.

References

Chapter 1

1. Ira N. Levine, *Physical Chemistry*, McGraw-Hill, New York (1978).
2. D. E. Goldsack, R. Franchetto, and A. Franchetto, Solvation effects on the conductivity of concentrated electrolyte solutions, *Can. J. Chem.* **54**, 2953–2966 (1976).
3. International Union of Pure and Applied Chemistry. Manual of Symbols and Terminology for Physicochemical Quantities and Units: Appendix III, Electrochemical nomenclature, *Pure and Appl. Chem.* **37**, 503–516 (1974).
4. H. Lamb, *Hydrodynamics*, Dover, New York (1949).

Chapter 2

1. R. M. Fuoss, The conductance–concentration function for alkali halides in dioxane–water mixtures, *Rev. Pure Appl. Chem.* **18**, 125–136 (1968).
2. F. Kohlrausch and A. Heydweiller, Über reines Wasser, *Z. Phys. Chem.* **14**, 317 (1894).
3. L. Onsager, The theory of electrolytes (I), *Physik. Z.* **27**, 388–392 (1926). The theory of electrolytes (II), *Physik. Z.* **28**, 277–298 (1927).
4. (a) J. Barthel, Conductance of electrolyte solutions, *Angew. Chem.* **7**(4), 260–277 (1968); (b) H. Falkenhagen, W. Ebeling, and W. D. Kraeft in *Ionic Interactions* (S. Petrucci, ed.), Academic Press, New York (1971), pp. 62–116; (c) R. M. Fuoss, in *Chemical Physics of Ionic Solutions* (B. E. Conway and R. G. Barradas, eds.), John Wiley & Sons New York (1966), pp. 463–485; (d) H. S. Harned and B. B. Owen, *The Physical Chemistry of Electrolyte Solutions*, Reinhold, New York (1958); (e) Ion–Ion and Ion–Solvent Interactions, *Faraday Discussions* of the Chemical Society No. 64 (1977); (f) R. Fernandez-Prini, in *Physical Chemistry of Organic Solvent Systems* (A. K. Coving and T. Dickinson, eds.), Plenum, London (1973), pp. 525–614.
5. R. M. Fuoss and F. Accascina, *Electrolytic Conductance*, Interscience, New York (1959).
6. M-S. Chen, Compatibility of conductance equations with Onsager's reciprocal relation, *J. Phys. Chem.* **81**, 2022–2023 (1977).
7. H. Falkenhagen, M. Leist, and G. Kelbg, Theory of the conductivity of strong, nonassociative electrolytes at higher concentrations, *Ann. Phys.* (*Leipzig*) **11**, 51–59 (1952).
8. E. Pitts, An extension of the theory of the conductivity and viscosity of electrolyte solutions, *Proc. Roy. Soc. London Ser. A*, **217**, 43–70 (1953).

9. R. M. Fuoss and L. Onsager, Conductance of unassociated electrolytes, *J. Phys. Chem.* **61**, 668–682 (1957).

10. P. C. Carman, Debye-Hückel relaxation effects in conductance of electrolyte solutions: Trends at higher concentrations, *J. S. Afr. Chem. Inst.*, **XXVII**, 80–103 (1975).

11. R. M. Fuoss, Conductance–concentration function for the paired ion model, *J. Phys. Chem.* **82**, 2427–2440 (1978).

12. M-S. Chen, The conductance of mixed strong electrolytes at finite concentration, Ph.D. Thesis, Yale University (1969).

13. P. C. Carman, Logarithmic term in conductivity equation for dilute solutions of strong electrolytes, *J. Phys. Chem.* **74**, 1653–1654 (1970).

14. R. Fernandez-Prini and J. E. Prue, A comparison of conductance equations for unassociated electrolytes, *Z. Phys. Chem.* (*Leipzig*) **228**, 373–379 (1965).

15. J.-C. Justice, An interpretation for the distance parameter of the Fuoss–Onsager conductance equation in the case of ionic association, *Electrochim. Acta* **16**, 701–711 (1971).

16. E. M. Hanna, A. D. Pethybridge, and J. E. Prue, Ion association and the analysis of precise conductimetric data, *Electrochim. Acta.* **16**, 677–686 (1971).

17. R. M. Fuoss, L. Onsager, and J. F. Skinner, The conductance of symmetrical electrolytes. V. The conductance equation, *J. Phys. Chem.* **69**, 2581–2594 (1965).

18. R. M. Fuoss and K.-L. Hsia, Association of 1–1 salts in water, *Proc. Nat. Acad. Sci. USA.* **57**, 1550–1557 (1967).

19. R. Fernandez-Prini, Conductance of electrolyte solutions: A modified expression for its concentration dependence, *Trans. Faraday Soc.* **65**, 3311–3313 (1969).

20. J. Barthel, J.-C. Justice, and R. Wachter, Untersuchung der elektrischen Leitfähigkeit alkoholischer alkalialkohatlösungen VII. Discussion of the distance parameters with the help of the extended conductance equations, *Z. Phys. Chem.* (*Frankfurt am Main*) **84**, 100–113 (1973).

21. R. M. Fuoss, Ionic association. III. The equilibrium between ion pairs and free ions, *J. Amer. Chem. Soc.* **80**, 5059–5061 (1958).

22. R. M. Fuoss, Derivation of the Justice conductance equation, *J. Phys. Chem.* **78**, 1383–1386 (1974).

23. P. Beronius, Remarks on the problem of finding best set conductance parameters for electrolyte solutions, *Acta Chem. Scand. Ser. A* **30**, 115–120 (1976).

24. S. D. Hamann, *Physico-Chemical Effects of Pressure*, Butterworth's Scientific Publications, London (1957).

25. S. B. Brummer and A. B. Gancy, in *Water and Aqueous Solutions* (R. A. Horne, ed.), Wiley-Interscience, New York (1972), pp. 745–770.

26. (a) A. B. Gancy and S. B. Brummer, Conductance of aqueous electrolyte solutions at high pressures, *J. Chem. and Eng. Data* **16**, 385–388 (1971); (b) A. B. Gancy and S. B. Brummer, The effect of solution concentration on the high-pressure coefficient of ionic conductance, *J. Phys. Chem.* **73**, 2429–2436 (1969).

27. M. Nakahara, K. Shimizu, and J. Osugi, Ionic solutions under high pressures, III. Pressure and temperature effects on the mobilities of K^+ and Cl^- ions, *Rev. Phys. Chem. Japan* **42**, 12–25 (1972).

28. R. A. Robinson and R. H. Stokes, Variation of equivalent conductance with concentration and temperature, *J. Amer. Chem. Soc.* **76**, 1991–1994 (1954).

29. R. M. Fuoss and T. Shedlovsky, Extrapolation of conductance data for weak electrolytes, *J. Amer. Chem. Soc.* **71**, 1496–1498 (1949).

30. J. F. Cukurins and W. Strauss, Effect of pressures to 3 kbar on the electrical conductivity of tetramethylammonium bromide and hydrobromic acid in propan-1-ol and propan-2-ol, *Aust. J. Chem.* **29**, 249–256 (1976).

31. W. A. Adams and K. J. Laidler, Electrical conductivities of quaternary ammonium salts in acetone: Part I. Pressure and temperature effects. Part II. The mechanisms of transport, *Can. J. Chem.* **46**, 1977–2011 (1968).

32. J. F. Skinner and R. M. Fuoss, Effect of pressure on conductance: II. Walden products and ionic association in methanol, *J. Phys. Chem.* **70**, 1426–1433 (1966).

33. E. Inada, The pressure and temperature effects on the Walden products of Na^+, K^+, Cs^+, and I^- in water, *Rev. Phys. Chem. Jap.* **46**, 19–30 (1976).

34. R. M. Fuoss and C. A. Kraus, Properties of electrolytic solutions: XV. Thermodynamic properties of very weak electrolytes, *J. Amer. Chem. Soc.* **57**, 1–4 (1935).

35. A. H. Ewald and J. A. Scudder, The effect of pressure on the conductance of some iodides in acetone and 2-methylpropan-1-ol, *Aust. J. Chem.* **23**, 1939–1945 (1970).

36. S. L. Wellington, The effect of pressure and temperature on ionic conductance in 2-propanol, *Diss. Abstr. Int. B*, **33**, 4751 (1973).

37. A. S. Quist and W. L. Marshall, Electrical conductances of aqueous sodium chloride solutions from 0 to 800° and at pressures to 4000 bars, *J. Phys. Chem.* **72**, 684–703 (1968).

38. (a) A. S. Quist, E. U. Franck, H. R. Jolley, and W. L. Marshall, Electrical conductances of aqueous solutions at high temperature and pressure: I. The conductances of potassium sulfate–water solutions from 25 to 800° and at pressures up to 4000 bars, *J. Phys. Chem.* **67**, 2453–2459 (1963); (b) A. S. Quist, W. L. Marshall, and H. R. Jolley, Electrical conductances of aqueous solutions at high temperature and pressure: II. The conductances and ionization constants of sulfuric acid–water solutions from 0 to 800° and at pressures up to 4000 bars, *J. Phys. Chem.* **69**, 2726–2735 (1965); (c) A. S. Quist and W. L. Marshall, Electrical conductances of aqueous solutions at high temperature and pressure: III. The conductances of potassium bisulfate solutions from 0 to 700° and at pressures to 4000 bars, *J. Phys. Chem.* **70**, 3714–3725 (1966); (d) A. S. Quist and W. L. Marshall, Electrical conductances of aqueous sodium bromide solutions from 0 to 800° and at pressures to 4000 bars, *J. Phys. Chem.* **72**, 2100–2105 (1968); (e) L. A. Dunn and W. L. Marshall, Electrical conductances of aqueous sodium iodide and the comparative thermodynamic behaviour of aqueous sodium halide solutions to 800° and 4000 bars, *J. Phys. Chem.* **73**, 723–728 (1969); (f) A. S. Quist and W. L. Marshall, Electrical conductances of aqueous hydrogen bromide solutions from 0 to 800° and at pressures to 4000 bars, *J. Phys. Chem.* **72**, 1545–1552 (1968); (g) A. S. Quist and W. L. Marshall, Ionization equilibria in ammonia–water solutions to 700° and to 4000 bars of pressure, *J. Phys. Chem.* **72**, 3122–3128 (1968).

39. (a) E. U. Franck, Hochverdichteter Wasserdampf: I. Elektrolytische Leitfahigkeit in $KCl–H_2O$-losungen bis 750°C, *Z. Phys. Chem.* **8**, 92–106 (1956); (b) E. U. Franck, Hochverdichteter Wasserdampf: II. Ionendissoziation von KCl in H_2O bis 750°C, *Z. Phys. Chem.* **8**, 107–126 (1956).

40. K. von Mangold and E. U. Franck, Electrical conductivity of aqueous solutions at high temperatures and pressures: II. Alkali metal chloride in water up to 1000°C and 12 kbar, *Ber. Bunsenges Phys. Chem.* **73**, 21–27 (1969).

41. G. von Ritzert and E. U. Franck, Electrical conductivity of aqueous solutions at high temperatures and pressures: I. Electrolytic conductivity of potassium chloride, barium chloride, barium hydroxide and magnesium sulphate up to 750°C and 6 kbar, *Ber. Bunsenges Phys. Chem.* **72**, 798–808 (1968).

42. E. G. Larionov and P. A. Kryukov, The limiting equivalent conductivity of KCl at temperatures up to 150°C and pressures up to $8000/kg/cm^2$, International Conference on High Temperature and High Pressure Electrochemistry in Aqueous Solutions, University of Surrey (1973).

43. L. B. Yeatts, L. A. Dunn, and W. L. Marshall, Electrical conductances and ionization

behaviour of sodium chloride in dioxane–water solutions at 100° and pressures to 4000 bars, *J. Phys. Chem.* **75**, 1099–1106 (1971).

44. W. R. Gilkerson, The importance of the effect of the solvent dielectric constant on ion-pair formation in water at high temperatures and pressures, *J. Phys. Chem.* **74**, 746–750 (1970).

45. R. A Matheson, The thermodynamics of electrolyte equilibria in media of variable water concentration, *J. Phys. Chem.* **73**, 3635–3642 (1969).

46. R. L. Kay, D. F. Evans, and Sister M. A. Matesich, in *Solute–Solvent Interact* (J. F. Coetzee and J. F. Ritchie, eds.), Vol. 2, Marcel Dekker, New York (1976), pp. 105–153.

47. (a) M. Ueno, M. Nakahara, and J. Osugi, The electrical conductivities of NH_4Cl and $Me_nNH_{4-n}Cl$ ($n = 1, 2, 3$) in water, *Rev. Phys. Chem. Jap.* **45**, 9–16 (1975); (b) M. Ueno, M. Nakahara, and J. Osugi, The electrical conductivities of $Et_nNH_{4-n}Cl$ ($n = 1, 2, 3$) in water, *Rev. Phys. Chem. Jap.* **45**, 17–22 (1975).

48. K. E. Bett and J. B. Cappi, Effect of pressure on the viscosity of water, *Nature* **207**, 620 (1965).

49. T. S. Burn and H. Høiland, Pressure dependence of the electric conductance of aqueous solutions of LiCl, KCl and $(CH_3)_4NCl$. Correlation of kinetic parameters, *Electrochim. Acta*, **21**, 51–57 (1976).

50. R. A. Horne and R. P. Young, The electrical conductivity of aqueous tetraalkylammonium halide solutions under hydrostatic pressure, *J. Phys. Chem.* **72**, 1763–1767 (1968).

51. M. Nakahara, Ionic solutions under high pressures: IV. Effects of pressure on the mobilities and hydration of Bu_4N^+, Me_4N^+, K^+ and Cl^- ions, *Rev. Phys. Chem. Jap.* **42**, 75–84 (1972).

52. F. H. Fisher, The effect of pressure on the equilibrium of magnesium sulfate, *J. Phys. Chem.* **65**, 1607–1611 (1962).

53. F. H. Fisher and D. F. Davis, The effect of pressure on the dissociation of manganese sulfate ion pairs in water, *J. Phys. Chem.* **69**, 2595–2598 (1965).

54. F. H. Fisher and A. P. Fox, KSO_4^-, $NaSO_4^-$ and $MgCl^+$ ion pairs in aqueous solutions up to 2000 atm, *J. Solution Chem.* **6**, 641–650 (1977).

55. F. H. Fisher and A. P. Fox, $LiSO_4^-$, $RbSO_4^-$, $CsSO_4^-$ and $(NH_4)SO_4^-$ ion pairs in aqueous solutions at pressures up to 2000 atm, *J. Solution Chem.* **7**, 561–570 (1978).

56. S. D. Hamann, P. J. Pearce, and W. Strauss, The effect of pressure on the dissociation on lanthanum ferricyanide ion pairs in water, *J. Phys. Chem.* **68**, 375–380 (1964).

57 J. Osugi, K. Shimizu, M. Nakahara, E. Hirayama, Y. Matsubara, and M. Ueno, Aqueous ionic solutions at high pressures, Proceedings 4th International Conference on High Pressures, 610–614 (1974).

58. P. J. Pearce and W. Strauss, Conductances of potassium chloride and tetrabutylammonium picrate in dioxane–water mixtures at pressures up to 2500 bars, *Aust. J. Chem.* **23**, 905–913 (1970).

59. S. B. Brummer and G. J. Hills, Kinetics of ionic conductance: Part 2. Temperature and pressure coefficients of conductance, *Trans. Faraday Soc.* **57**, 1823–1837 (1961).

60. J. F. Skinner and R. M. Fuoss, Effect of pressure on conductance: II. Walden products and ionic association in methanol, *J. Phys. Chem.* **70**, 1426–1433 (1966).

61. E. L. Cussler and R. M. Fuoss, Effect of pressure on conductance: IV. Ionic association and Walden products in ethanol, *J. Phys. Chem.* **71**, 4459–4463 (1967).

62. S. B. Brummer, Temperature and pressure coefficients of ionic conductance in N,N-dimethylformamide, *J. Chem. Phys.* **42**, 1636–1646 (1965).

63. F. Barreira and G. J. Hills, Kinetics of ionic migration, Part 3: Pressure and temperature coefficients of conductance in nitrobenzene, *Trans. Faraday Soc.* **64**, 1359–1375 (1968).

64. J. F. Skinner and R. M. Fuoss, Effect of pressure on conductance: I. Tetraisoamylammonium picrate in diethyl ether and in benzene, *J. Phys. Chem.* **69**, 1437–1443 (1965).

65. M. Eigen and L. De Maeyer, Self-dissociation and protonic charge transport in water and ice, *Proc. Roy. Soc. London*, **A247**, 505–533 (1958).
66. D. A. Lown and H. R. Thirsk, Proton transfer conductance in aqueous solution: Part 1. Conductance of concentrated aqueous alkali metal hydroxide solutions at elevated temperatures and pressures. Part 2. Effect of pressure on the electrical conductivity of concentrated orthophosphoric acid in water at 25°C, *Trans. Faraday Soc.* **67**, 132–152 (1971).
67. M. Nakahara and J. Osugi, Ionic solutions under high pressures: VII. Mobility of hydrogen ion in water at 15, 25 and 40°C up to 5000 atm, *Rev. Phys. Chem. Jap.*, **47**, 1–11 (1977).
68. K. Todheide, in *Water, A Comprehensive Treatise* (F. Franks, ed.) Vol. 1, Plenum Press, New York (1972), pp. 463–514.
69. M. Goffredi and T. Shedlovsky, Studies of electrolytic conductance in alcohol–water mixtures: IV. Hydrochloric acid in 1-propanol-water mixtures at 15, 25 and 35°, *J. Phys. Chem.* **71**, 2182–2186 (1967).
70. S. B. Brummer and G. J. Hills, Kinetics of ionic conductance: Part 1. Energies of activation and the constant volume principle, *Trans. Faraday Soc.* **57**, 1816–1822 (1971).
71. G. J. Hills, in *Chemical Physics of Ionic Solutions* (B. J. Conway and R. G. Barradas, eds.), pp. 521–539, John Wiley & Sons, New York (1966).
72. R. L. Kay, in *Water, A Comprehensive Treatise* (F. Franks, ed.), Vol. 3, Plenum Press, New York (1973), pp. 173–209.
73. von M. Born, Uber die Beweglichkeit der elektrolytischen Ionen, *Z. fuer Physik* **1**, 221–249 (1920).
74. R. M. Fuoss, Dependence of the Walden product on the dielectric constant, *Proc. Nat. Acad. Sci. U.S.A.* **45**, 807–813 (1959).
75. R. H. Boyd, Letters to the Editor, *J. Chem. Phys.* **39**, 2376 (1963).
76. R. Zwanzig, Dielectric friction on a moving ion: II. Revised theory, *J. Chem. Phys.* **52**, 3625–3628 (1970).
77. T. L. Broadwater, T. J. Murphy, and D. F. Evans, Conductance of binary asymmetric electrolytes in methanol, *J. Phys. Chem.* **80**, 753–757 (1976).
78. R. Fernandez-Prini, Ionic conductivities and dielectric friction, *J. Phys. Chem.* **77**, 1314–1315 (1973).
79. (a) J. Hubbard and L. Onsager, Dielectric dispersion and dielectric friction in electrolyte solutions, I, *J. Chem. Phys.* **67**, 4850–4857 (1977); (b) J. B. Hubbard, Dielectric dispersion and dielectric friction in electrolyte solutions, II, *J. Chem. Phys.* **68**, 1649–1664 (1978).
80. D. F. Evans, C. Chan, and B. C. Lamartine, The effect of charge upon mobility. A critical examination of the Zwanzig equation, *J. Amer. Chem. Soc.* **99**, 6492–6496 (1977).
81. H. S. Frank and W-Y Wen, Ion–solvent interaction III. Structural aspects of ion–solvent interaction in aqueous solutions: A suggested picture of water structure, *Discuss. Faraday Soc.* **24**, 133–140 (1957).
82. A. Geiger and H. G. Hertz, Proton magnetic relaxation study of water orientation around I^- and Li^+, *J. Solution Chem.* **5**, 365–388 (1976).
83. *Water, A Comprehensive Treatise* (F. Franks, ed.), Vols. 1 and 3, Plenum Press, New York (1973).
84. T. L. Broadwater and R. L. Kay, The temperature coefficient of conductance for the alkali metal, halide, tetraalkylammonium, halate and perhalate ions in D_2O, *J. Solution Chem.* **4**, 745–762 (1975).
85. M. Nakahara and J. Osugi, Ionic solutions under high pressure: VI. Mobilities and hydration of monocarboxylate ions, *Rev. Phys. Chem. Jap.* **45**, 1–8 (1975).
86. F. J. Millero, in *Water and Aqueous Solutions* (R. A. Horne, ed.), John Wiley & Sons, New York (1972), pp. 519–564.

87. R. A. Horne, in *Advances in High Pressure Research* (R. S. Bradley, ed.), Vol. 2, Academic Press, London (1969), pp. 169–223.

88. Y. Lee and J. Jonas, Effect of pressure on proton spin–lattice relaxation in several concentrated aqueous electrolyte solutions, *J. Magnetic Resonance* 5, 267–272 (1971).

89. Y. K. Lee, J. H. Campbell and J. Jonas, Effect of pressure on deuteron spin–lattice relaxation in several concentrated deuterium oxide diamagnetic electrolyte solutions, *J. Chem. Phys.* 60, 3537–3543 (1974).

90. R. A. Robinson and R. H. Stokes, *Electrolyte Solutions*, 2nd ed. rev., Butterworths, London (1959), Chapter 6.

91. B. S. Smolyakov, Limiting equivalent ionic conductance up to 200°C, International Conference on High Temperature and High Pressure Electrochemistry in Aqueous Solutions, University of Surrey (1973).

92. V. M. Valyashko, Studies of water–salt systems at elevated temperatures and pressures, *Ber. Bunsenges. Phys. Chem.* 81, 388–396 (1977).

93. C. Shin and C. M. Criss, Partial molal heat capacities of tetraalkylammonium bromides in methanol from 10 to 50°C, *J. Solution Chem.* 7, 205–217 (1978).

94. C. Shin, I. Worsley, and C. M. Criss, Partial molal heat capacities of aqueous tetraalkylammonium bromides as functions of temperature, *J. Solution Chem.* 5, 867–879 (1976).

95. A. J. Pasztor and C. M. Criss, Apparent molal volumes and heat capacities of some 1:1 electrolytes in anhydrous methanol at 25°C, *J. Solution Chem.* 7, 27–43 (1978).

96. B. Watson and R. L. Kay, see Ref. 72, p. 201.

97. R. L. Kay and T. L. Broadwater, Solvent structure in aqueous mixtures: III. Ionic conductances in ethanol–water mixtures at 10 and 25°C., *J. Solution Chem.* 5, 57–76 (1976).

98. T. L. Broadwater and R. L. Kay, Solvent structure in aqueous mixtures: II. Ionic mobilities in tert-butyl alcohol–water mixtures at 25°C, *J. Phys. Chem.* 74, 3802–3812 (1970).

99. R. L. Kay and T. L. Broadwater, Solvent structure in aqueous mixtures: I. Ionic mixtures in dioxane–water mixtures at 25°C, *Electrochim. Acta* 16, 667–676 (1971).

100. J. O'M. Bockris, J. A. Kitchener, S. Ignatowicz, and J. W. Tomlinson, Electric conductance in liquid silicates, *Trans. Faraday Soc.* 48, 75–91 (1952).

101. E. G. Larionov, Calculation of maximum electric conductivity of individual ions at high temperatures and pressures, *Otdelenie Khimicheskikh Nauk*, Akademiia Nauk, Novosibirsk, SSSR (1977), pp. 10–19.

102. M.-C. and J.-C. Justice and R. Bury, *Electrochim. Acta* 16, 687–700 (1971).

103. D. F. Evans, C. Zawoski, and R. L. Kay, *J. Phys. Chem.* 69, 3878–3885 (1965).

104. E. M. Hanna, A. D. Pethybridge, and J. E. Prue, *Electrochim. Acta.* 16, 677–686 (1971).

105. P. C. Carman, *J. Solution Chem.* 6, 609–624 (1977).

106. R. A. Robinson and R. H. Stokes, *Electrolyte Solutions*, Butterworths, London (1966).

107. H. Falkenhagen, W. Ebeling, and H. G. Hertz, *Theorie der Elektrolyte*, Hirzel, Leipzig (1970).

Chapter 3

1. H. S. Harned and B. B. Owen, *The Physical Chemistry of Electrolytic Solutions*, Reinhold, New York (1943).

2. R. A. Robinson and R. H. Stokes, *Electrolyte Solutions*, Butterworths, London (1966).

3. R. Fuoss and L. Onsager, Conductance of unassociated electrolytes, *J. Phys. Chem.* 61, 668–682 (1957).

4. R. M Fuoss, Conductance–concentration function for the paired ion model, *J. Phys. Chem.* **82**, 2427–2440 (1978).

5. D. E. Goldsack, R. Franchetto, and A. Franchetto, Solvation effects on the conductivity of concentrated electrolyte solutions, *Can. J. Chem.* **54**, 2953–2966 (1976).

6. A. B. Gancy and S. B. Brummer, The effect of solution concentration on the high-pressure coefficient of ionic conductance, *J. Phys. Chem.* **73**, 2429–2436 (1969).

7. T. S. Brun and H. Høiland, Pressure dependence of the electric conductance of aqueous solutions of LiCl, KCl and $(CH_3)_4NCl$, *Electrochim. Acta* **21**, 51–57 (1976).

8. J. U. Hwang, H. D. Ludemann, and D. Hartmann, Die elektrische Leitfahigkeit konzentrierter wassriger Alkalihalogenidlosungen bei hohen Drucken und Temperaturen, *High Temperatures–High Pressures* **2**, 651–669 (1970).

9. D. A. Lown and H. R. Thirsk, Proton transfer conductance in aqueous solution, *Trans. Faraday Soc.* **67**, 132–152 (1971).

10. S. K. Fellows, Conductance measurements on aqueous solutions over extended ranges of concentration and temperature, Ph.D. thesis, Victoria University of Wellington, Wellington, New Zealand (1971).

11. L. F. Grantham and S. J. Yosim, Negative temperature coefficients of electrical conductance in molten salts, *J. Chem. Phys.* **45**, 1192–1198 (1966).

12. C. A. Angell and R. D. Bressel, Fluidity and conductance in aqueous electrolyte solutions, I, *J. Phys. Chem.* **76**, 3244–3253 (1972).

13. C. A. Angell and E. J. Sare, Glass-forming composition regions and glass transition temperatures for aqueous electrolyte solutions, *J. Chem. Phys.* **52**, 1058–1068 (1970).

14. B. F. Wishaw and R. H. Stokes, The diffusion coefficients and conductances of some concentrated electrolyte solutions at 25°C, *J. Amer. Chem. Soc.* **76**, 2065–2071 (1954).

15. H. Falkenhagen, M. Leist, and G. Kelbg, Theory of the conductivity of strong, nonassociative electrolytes at higher concentrations, *Ann. Physik*, **6**(II), 51–59 (1952).

16. N-P. Yao and D. N. Bennion, Transport behaviour in dimethyl sulfoxide, III: Conductance–viscosity behaviour of tetra-N-amylammonium thiocyanate from infinite dilution to molten salt at 55°C, *J. Phys. Chem.* **75**, 3586–3591 (1971).

17. R. P. T. Tomkins, E. Andalaft, and G. J. Janz, Conductance, density and viscosity of NaI in anhydrous acetonitrile at 25°C, *Trans. Faraday Soc.* **65**, 1906–1911 (1969).

18. L. C. Kenausis, E. C. Evers, and C. A. Kraus, Mechanisms underlying the equilibrium reactions between ions and ion pairs in solutions of electrolytes, I, *Proc. Nat. Acad. Sci. U.S.* **48**, 121–128 (1962).

19. M. H. Cohen and D. Turnbull, Molecular transport in liquid and glasses, *J. Chem. Phys.* **31**, 1164–1169 (1959).

20. G. Adam and J. H. Gibbs, On the temperature dependence of cooperative relaxation properties in glass-forming liquids, *J. Chem. Phys.* **43**, 139–146 (1965).

21. C. A. Angell, Free volume model for transport in fused salts: Electrical conductance in glass-forming nitrate melts, *J. Phys. Chem.* **68**, 1917–1919 (1964).

22. C. A. Angell, Free volume-entropy interpretation of the electrical conductance of aqueous electrolyte solutions in the concentration range 2–20 N, *J. Phys. Chem.* **70**, 3988–3997 (1966).

23. C. T. Moynihan, C. R. Smalley, C. A. Angell, and E. J. Sare, Conductance, viscosity, density, proton magnetic resonance spectra and glass transition temperatures of calcium nitrate tetrahydrate–cadmium nitrate tetrahydrate melts, *J. Phys. Chem.* **73**, 2287–2293 (1969).

24. N. Islam and K. Ismail, Temperature and composition dependence of transport properties of cobalt (II) chloride–calcium nitrate tetrahydrate melts, *J. Phys. Chem.* **80**, 1929–1933 (1976).

25. C. T. Moynihan, The temperature dependence of transport properties of ionic liquids.

The conductance and viscosity of calcium nitrate tetrahydrate and sodium thiosulfate pentahydrate, *J. Phys. Chem.* **70**, 3399–3403 (1966).

26. C. A. Angell, On the importance of the metastable liquid state and glass transition phenomenon to transport and structure studies in ionic liquids: I. Transport properties, *J. Phys. Chem.* **70**, 2793–2803 (1966).

27. S. I. Smedley and I. Torrie, Transport in molten $CaCl_2 \cdot 5 \cdot 99, 5 \cdot 33H_2O$ under pressure, *J. Phys. Chem.* **82**, 238–241 (1978).

28. (a) T. G. Chang and D. E. Irish, Raman and infrared spectral study of magnesium nitrate–water systems, *J. Phys. Chem.* **77**, 52–57 (1973); (b) M. Peleg, A Raman spectroscopic investigation of the magnesium nitrate–water system, *J. Phys. Chem.* **76**, 1019–1025 (1972); (c) R. F. Hester and R. A. Plane, Raman spectrophotometric study of complex formation in aqueous solutions of calcium nitrate, *J. Chem. Phys.* **40**, 411–414 (1964).

29. E. J. Sare, C. T. Moynihan, and C. A. Angell, Proton magnetic resonance chemical shifts and the hydrogen bond in concentrated aqueous electrolyte solutions, *J. Phys. Chem.* **77**, 1869–1876 (1973).

30. P. S. Leung and G. J. Safford, A neutron inelastic scattering investigation of the diffusion kinetics of H_2O molecules and hydration complexes in concentrated ionic solutions, *J. Solution Chem.* **2**, 525–547 (1973).

31. W. Kauzmann, The nature of the glassy state and the behavior of liquids at low temperatures, *Chem. Rev.* **43**, 219–256 (1948).

32. C. A. Angell and C. T. Moynihan, in *Molten Salts* (G. Mamontov, ed.), Marcel Dekker, New York (1969), p. 315.

33. C. A. Angell and J. C. Tucker, Heat capacities and fusion entropies of the tetrahydrates of calcium nitrate, cadmium nitrate and magnesium acetate, *J. Phys. Chem.* **78**, 278–281 (1974).

34. C. A. Angell and C. T. Moynihan, in *Molten Salts* (G. Mamontov, ed.), Marcel Dekker, New York (1969), p. 346.

35. C. A. Angell and K. J. Rao, Configurational excitations in condensed matter, and the "bond lattice" model for the liquid-glass transition, *J. Chem. Phys.* **57**, 470–481 (1972).

36. L. Pickston, S. I. Smedley, and G. Woodall, The compressibility and electrical conductivity of concentrated aqueous calcium nitrate solutions to 6 kbar and 150°C, *J. Phys. Chem.* **81**, 581–586 (1977).

37. C. A. Angell, L. J. Pollard, and W. Strauss, Transport in ionic liquids under pressure, I: Concentrated calcium nitrate–water and magnesium chloride–water solutions, *J. Solution Chem.* **1**, 517–530 (1972).

38. C. A. Angell, L. J. Pollard, and W. Strauss, Transport in molten salts under pressure, I: Glass-forming nitrate melts, *J. Chem Phys.* **50**, 2694–2705 (1969).

39. A. F. M. Barton and R. J. Speedy, Simultaneous conductance and volume measurements on molten salts at high pressure, *J. Chem. Soc., Faraday Trans. I* **70**, 506–527 (1974).

40. R. O. Davies and G. O. Jones, Thermodynamics and kinetic properties of glasses, *Adv. Phys.* **2**, 370–410 (1953).

41. M. Goldstein, Some thermodynamic aspects of the glass transition: Free volume, entropy and enthalpy theories, *J. Chem. Phys.* **39**, 3369–3374 (1963).

42. C. A. Angell and W. Sichina, Thermodynamics of the glass transition: Empirical aspects; Part II. The glass transition: thermodynamic and kinetic aspects, *Ann. N.Y. Acad. Sci.* **279**, 53–67 (1976).

43. F. S. Howell, R. A. Bose, P. B. Macedo, and C. T. Moynihan, Electrical relaxation in a glass-forming molten salt, *J. Phys. Chem.* **78**, 639–648 (1974).

44. C. A. Angell, Oxide glasses in light of the "ideal glass" concept: II, Interpretations by reference to simple ionic glass behavior, *J. Amer. Ceramic Soc.* **51**, 125–134 (1968).

45. J. H. Ambrus, C. T. Moynihan, and P. B. Macedo, Conductivity relaxation in a concentrated aqueous electrolyte solution, *J. Phys. Chem.* **76**, 3287–3294 (1972).

46. C. T. Moynihan, N. Balitactac, L. Boone, and T. A. Litovitz, Comparison of shear and conductivity relaxation times for concentrated lithium chloride solutions, *J. Chem. Phys.* **55**, 3013–3019 (1971).

47. H. Tweer, J. H. Simmons, and P. B. Macedo, Application of the environmental relaxation model to the temperature dependence of the viscosity, *J. Chem. Phys.* **54**, 1952–1959 (1971).

Chapter 4

1. Milton Blander, ed., *Molten Salt Chemistry*, Interscience, New York (1964).

2. B. R. Sundheim, ed., *Fused Salts*, McGraw-Hill, New York (1964).

3. G. Mamontov, ed., *Molten Salts*, Marcel Dekker, New York (1969).

4. J. L. Copeland, *Transport Properties of Ionic Liquids*, Gordon and Breach Science, New York (1974).

5. C. T. Moynihan, in *Mass Transport in Fused Salts* (S. Petrucci, ed.), Vol. 1, Ionic Interactions, Academic Press, New York (1971), Chapter 5.

6. G. J. Janz, C. Solomons, and H. J. Gardner, Physical properties and constitution of molten salts, *Chem. Revs.* **58**, 461–508 (1958).

7. G. J. Janz and R. D. Reeves, Molten-salt electrolytes, *Adv. Electrochem. Electrochem. Eng.* **5**, 137–204 (1967).

8. D. Inman, A. D. Graves, and R. S. Sethi, Electrochemistry of molten salts, *Electrochemistry* **1**, 166–222 (1968–1969).

9. (a) D. Inman, A. D. Graves, and A. A. Nobile, Electrochemistry of molten salts, *Electrochemistry* **2**, 61–116 (1972); (b) D. Inman, J. E. Bowling, D. G. Lovering, and S. H. White, Electrochemistry of molten salts, *Electrochemistry* **4**, 78–166 (1974).

10. J. W. Tomlinson, in *Electrochemistry: The Past Thirty and the Next Thirty Years* (H. Bloom and F. Gutmann, eds.), Plenum, New York (1977), pp. 335–350.

11. K. Tödheide, Molten salts at high temperatures and high pressures, Proceedings International Symposium of Molten Salts, Washington, D.C. (1976), pp. 20–35.

12. C. J. Janz, F. W. Dampier, and P. K. Lorentz, Molten Salts: Electrical conductance density and viscosity data, Rensselaer Polytechnic Institute, Troy, N.Y. (1966).

13. B. Cleaver, S. I. Smedley, and P. N. Spencer, Effect of pressure on electrical conductivities of fused alkali metal halides and silver halides, *J. Chem. Soc., Faraday Trans. I* **68**, 1720–1734 (1972).

14. J. E. Bannard, A. F. M. Barton, and G. J. Hills, Transport parameters for the molar electrical conductivities of molten alkali nitrates at high pressures, *High Temperatures–High Pressures* **3**, 65–80 (1971).

15. R. Schamm and K. Tödheide, The electrolyte conductivity of molten sodium nitrite, potassium nitrite and sodium chlorate at pressures up to 6 kbar, *High Temperatures–High Pressures* **8**, 59–64 (1976).

16. B. Cleaver, S. I. Smedley, and P. N. Spencer, in *Atomic Transport in Solids and Liquids* (A. Lodding and T. Lagerwall, eds.), Verlag der Zeitschrift für Naturforschung, Tübingen (1971), pp. 355–359.

17. A. S. Quist, A. Würflinger, and K. Tödheide, Electrical conductivity of alkali nitrate molten salts at high pressure: 1. The pressure range up to 5.5 kbar, *Ber. Bunsenges Phys. Chem.* **76**, 652–661 (1972).

18. G. Schlichtharle, K. Tödheide, and E. U. Franck, Die elektrische Leitfahgkeit geschmolzener Alkalinitrate bei hohen Drucken: II. Der Druckbereich bis 12 kbar, *Ber. Bunsenges. Phys. Chem.* **76**, 1168–1175 (1972).

19. A. F. M. Barton and R. J. Speedy, Simultaneous conductance and volume measurements on molten salts at high pressure, *J. Chem. Soc. Faraday Trans. 1*, **70**, 506–527 (1974).

20. V. Pilz and K. Tödheide, Electrical conductance of molten alkali nitrates at high pressures: Pressures up to 55 kbars, *Ber. Bunsenges. Phys. Chem.* **77**, 29–36 (1973).

21. S. Glasstone, K. J. Laidler, and H. Eyring, *The Theory of Rate Processes*, McGraw-Hill, New York (1941).

22 J. O'M. Bockris, J. A. Kitchener, S. Ignatowicz, and J. W. Tomlinson, Electric conductance in liquid silicates, *Trans. Faraday Soc.* **48**, 75–91 (1952).

23. M. Goldstein, Viscous liquids and the glass transition: A potential energy barrier picture, *J. Chem. Phys.* **51**, 3728–3739 (1969).

24. B. J. Alder and T. Einwohner, Free-path distribution for hard spheres, *J. Chem. Phys.* **43**, 3399–3400 (1965).

25. S. I. Smedley and L. V. Woodcock, Kirkwood–Rice–Allnatt kinetic theory of transport in liquids, *J. Chem. Soc. Faraday Trans. II* **69**, 955–966 (1973).

26. (a) R. Fürth, The theory of the liquid state, *Proc. Cambridge Phil. Soc.* **37**, 252–280 (1941); (b) R. Fürth, The hole theory of viscous flow, *Proc. Cambridge Phil. Soc.* **37**, 281–240 (1941).

27. J. O'M. Bockris, E. H. Crook, H. Bloom, and N. E. Richards, The electric conductance of simple molten salts, *Proc. Roy. Soc. London* **A255**, 558–578 (1960).

28. T. Emi and J. O'M. Bockris, Semiempirical calculation of $3.7RT_m$ term in the heat of activation for viscous flow of ionic liquid, *J. Phys. Chem.* **74**, 159–163 (1970).

29. H. Eyring and M. S. Jhon, *Significant Liquid Structures*, John Wiley & Sons, New York (1969).

30. D. Henderson and W. Jost, in *Physical Chemistry, An Advanced Treatise* (H. Eyring, ed.), Vol. VIIIA, Academic Press, New York (1971), Chapter 5.

31. W-C. Lu, T. Ree, V. G. Gerrard, and H. Eyring, Significant-structure theory applied to molten salts, *J. Chem. Phys.* **49**, 797–804 (1968).

32. J. P. Hansen and I. R. McDonald, *Theory of Simple Liquids*, Academic Press, New York (1976).

33. B. Berne and S. A. Rice, On the kinetic theory of dense fluids: XVI. The ideal ionic melt, *J. Chem. Phys.* **40**, 1347–1362 (1964).

34. S. A. Rice and P. Gray, *The Statistical Mechanics of Simple Liquids*, Interscience, New York (1965).

35. J. G. Kirkwood, The statistical mechanical theory of transport processes: 1. General theory, *J. Chem. Phys.* **14**, 180–201 (1946).

36. (a) S. A. Rice and A. R. Allnatt, On the kinetic theory of dense fluids: VI. Singlet distribution function for rigid spheres with attractive potential, *J. Chem. Phys.* **34**, 2144–2155 (1961); (b) A. R. Allnatt and S. A. Rice, On the kinetic theory of dense fluids: VII. The doublet distribution function for rigid spheres with an attractive potential, *J. Chem. Phys.* **34**, 2156–2165 (1961).

37. S. A. Rice and P. Gray, *The Statistical Mechanics of Simple Liquids*, Interscience, New York (1965), p. 429.

38. E. Helfand, Theory of the molecular friction constant, *Phys. Fluids* **4**, 681–691 (1961).

39. G. Morrison and J. E. Lind, Jr., Friction constants for fused salts, *J. Phys. Chem.* **72**, 3001–3006 (1968).

40. K. Ichikawa and M. Shimoji, Self-diffusion and viscosity of molten alkali halides, *Trans. Faraday Soc.* **66**, 843–849 (1970).

41. S. A. Rice, Kinetic theory of ideal ionic melts, *Trans. Faraday Soc.* **58**, 499–510 (1962).

42. G. J. Janz, ed., *Molten Salts Handbook*, Academic Press, New York (1967).

43. J. O'M. Bockris, S. R. Richards, and L. Nanis, Self-diffusion and structure in molten Group II chlorides, *J. Phys. Chem.* **69**, 1627–1637 (1965).

44. R. A. Fisher and R. O. Watts, The friction coefficient formalism in the statistical mechanics of transport processes, *Aust. J. Phys.* **25**, 21–31 (1972).
45. G. Ciccotti, G. Jacucci, and I. R. McDonald, Transport properties of molten alkali halides, *Phys. Rev. A* **13**, 426–436 (1976).
46. J. P. Hansen and I. R. McDonald, Self-diffusion and electrical conductance in a simple molten salt, *J. Phys. C: Solid State Phys.* **7**, L384–L386 (1974).
47. J. P. Hansen and I. R. McDonald, Statistical mechanics of dense ionized matter: IV. Density and charge fluctuations in a simple molten salt, *Phys. Rev. A*, **11**, 2111–2123 (1975).
48. R. W. Laity, Formalisms and models for ionic transport, *Faraday Soc. Discuss.* **32**, 172–180 (1961).
49. R. W. Laity, General approach to the study of electrical conductance and its relation to mass transport phenomena, *J. Chem. Phys.* **30**, 682–691 (1959).
50. J. W. Tomlinson, in *Electrochemistry: The Past Thirty and the Next Thirty Years* (H. Bloom and F. Gutmann, eds.) Plenum, New York (1977), pp. 335–350.
51. M. J. Rice and W. Roth, Ionic transport in super ionic conductors. Theoretical model, *J. Solid State Chem.* **4**, 294–310 (1972).

Chapter 5

1. L. F. Grantham and S. J. Yosim, Negative temperature coefficients of electrical conductance in molten salts, *J. Phys. Chem.* **45**, 1192–1198 (1966).
2. L. F. Grantham, Electrical conductivities of molten cadmium–cadmium halide solutions, *J. Chem. Phys.* **44**, 1509–1513 (1966).
3. B. Bardoll and K. Tödheide, Die elektrische Leitfahigkeit und der Dissoziationzustand der geschmolzenen Quecksilberhalogenide bei Drücken bis 6 kbar, *Ber Bunsenges Phys. Chem.* **79**, 490–497 (1975).
4. von G. Treiber and K. Tödheide, Der kontinuierliche Ubergang vom Isolator zum Ionenleiter am Beispiel des Wismutchchlorids, *Ber Bunsenges Phys. Chem.* **77**, 541–547 (1973).
5. W. B. Holzapfel, Effect of pressure and temperature on the conductivity and ionic dissociation of water up to 100 kbar and 1000°C, *J. Chem. Phys.* **50**, 4424–4428 (1969).
6. D. J. Bearcroft and N. H. Nachtrieb, Electrical conductance of salts in liquid iodine; I: Iodide donor solutes, *J. Phys. Chem.* **71**, 316–323 (1967).
7. B. Cleaver and S. I. Smedley, Pressure dependence of electrical conductivity for fused mercuric halides, *Trans. Faraday Soc.* **67**, 1115–1127 (1971).
8. A. J. Darnell and W. A. McCollum, Fusion curve and electrical conductivity of molten $MgCl_2$ and HgI_2 at elevated pressure, *J. Chem. Phys.* **55**, 116–122 (1971).
9. J. E. Bannard and G. Treiber, The effect of temperature and pressure on the electrical conductance of molten mercury (II) iodide, *High Temperatures–High Pressures* **5**, 177–182 (1973).
10. B. Cleaver and P. N. Spencer, Isothermal compressibilities and thermal pressure coefficients of molten salts, *High Temperatures–High Pressures* **7**, 539–547 (1975).
11. B. Cleaver, P. N. Spencer, and M. A. Quddus, Effect of pressure on the electrical conductivities of some molten B group metal iodides and iodine, *J. Chem. Soc. Faraday I*, **3**, 686–696 (1978).
12. B. Cleaver and P. Zani, to be published; see Ref. 27 of Ref. 11.
13. K. Tödheide, in *Water, A Comprehensive Treatise* (F. Franks, ed.), Vol. 1, Plenum Press, New York (1972), Chapter 13.
14. M. Buback and E. U. Franck, Ionic conductivity of liquid NH_4Cl from triple point to critical point, *Ber. Bunsenges. Phys. Chem.* **77**, 1074–1079 (1973).

Chapter 6

1. D. Goldberg, ed., *The Sea*, Vol. 5, *Marine Chemistry*, Wiley–Interscience, New York (1974).
2. F. J. Millero, in *The Sea* (D. Goldberg, ed.), Vol. 5, Wiley–Interscience, New York (1974).
3. Ninth report of the joint panel on oceanographic tables and standards. UNESCO Technical Papers in Marine Science, No. 30, Paris, 11–13 September 1978, UNESCO (1979).
4. M-S. Chen and L. Onsager, The generalized conductance equation, *J. Phys. Chem.* **81**, 2017–2021 (1977).
5. R. A. Cox, in *Chemical Oceanography* (J. P. Riley and G. Skinrow, eds.), Academic Press, London (1965), pp. 73–120.
6. C. T. Chen and F. J. Millero, The specific volume of sea water at high pressures, *Deep-Sea Res.* **23**, 595–612 (1976).
7. D. N. Connors and P. K. Weyl, The partial equivalent conductance of salts in seawater and the density conductance relationship, *Limnol. Oceanog.* **13**, 35–50 (1968).
8. D. A. Lown and H. R. Thirsk, Proton transfer conductance in aqueous solution; Part 1, Conductance of concentrated aqueous alkali metal hydroxide solutions at elevated temperatures and pressures, *Trans. Faraday Soc.* **67**, 132–148 (1971).
9. S. K. Fellows, High temperature conductance of concentrated salt solutions, Ph.D. Thesis, Victoria University of Wellington, New Zealand (1971).
10. R. A. Horne and R. P. Young, The electrical conductivity of aqueous 0.03 to 4.0 M potassium chloride solutions under hydrostatic pressure, *J. Phys. Chem.* **71**, 3824–3832 (1967).
11. J. U. Hwang, H. D. Ludemann, and D. Hartmann, Die elektrische Leitfahigkeit konzentrierter wassriger Alkalihalogenidlosungen bei hohen Drucken und Temperaturen, *High Temperatures–High Pressures* **2**, 651–669 (1970).
12. A. Bradshaw and K. E. Schleicher, The effect of pressure on the electrical conductance of sea water, *Deep-Sea Res.* **12**, 151–162 (1965).
13. F. H. Fisher, Multistate dissociation and the effect of pressure on the equilibrium on magnesium sulfate, *J. Phys. Chem.* **69**, 695–698 (1965).
14. F. H. Fisher, and A. P. Fox, KSO_4^-, $NaSO_4^-$, and $MgCl^+$ ion pairs in aqueous solutions up to 2000 atm, *J. Solution Chem.* **6**, 641–650 (1977).
15. A. J. Ellis and W. A. J. Mahon, *Chemistry and Geothermal Systems*, Academic Press, New York (1977).
16. B. S. Smolyakov, Limiting Equivalent Ionic Conductance up to 200°C, International Conference on High Temperature and High Pressure Electrochemistry in Aqueous Solutions, University of Surrey (1973), pp. 177–181.
17. W. L. Marshall, Predictions of the geochemical behaviour of aqueous electrolytes at high temperatures and pressures, *Chemical Geology* **10**, 56–58 (1972).
18. A. J. Ellis and D. W. Anderson, The effect of pressure on the first acid dissociation constants of "sulphurous" and phosphoric acids, *J. Chem Soc.* **342**, 1765–1767 (1961).
19. A. J. Ellis and D. W. Anderson, The first acid dissociation constant of hydrogen sulphide at high pressures, *J. Chem. Soc.* **917**, 4678–4680 (1961).
20. (a) A. J. Ellis, The effect of pressure on the first dissociation constant of "carbonic acid," *J. Chem. Soc.* **750**, 3689–3699 (1959); (b) A. J. Read, The first ionization constant of carbonic acid from 25 to 250°C and to 2000 bar, *J. Solution Chem.* **4**, 53–70 (1975).
21. S. D. Hamann, *Physico-chemical Effects of Pressure*, Butterworths, London (1957).

22. R. W. Henly and A. McNabb, Magmatic vapor plumes and ground water interaction in porphyry copper emplacement, *Economic Geology* **73**, 1–19 (1978).
23. H. S. Waff, Theoretical consideration of electrical conductivity in a partially molten mantle and implications for geothermometry, *J. Geophys. Res.* **79**, 4003–4010 (1974).
24. J. O'M. Bockris, J. A. Kitchener, S. Ignatowicz, and J. W. Tomlinson, The electrical conductivity of silicate melts: Systems containing Ca, Mn and Al, *Discuss. Faraday Soc.* **4**, 265–281 (1948).
25. J. O'M. Bockris, J. A. Kitchener, S. Ignatowicz, and J. W. Tomlinson, Electric conductance in liquid silicates, *Trans. Faraday Soc.* **48**, 75–91 (1952).
26. J. O'M. Bockris, J. A. Kitchener, and A. E. Davies, Electric transport in liquid silicates, *Trans. Faraday Soc.* **48**, 536–548 (1951).
27. J. O'M. Bockris and G. W. Mellors, Electric conductance in liquid lead silicates and borates, *J. Phys. Chem.* **60**, 1321–1328 (1956).
28. (a) R. E. Tickle, The electrical conductance of molten alkali silicates, I, Experiments and results, *Phys. Chem. Glasses* **8**, 101–112 (1967); (b) R. E. Tickle, The electrical conductance of molten alkali silicates, II, Theoretical discussion, *Phys. Chem. Glasses* **8**, 113–124 (1967).
29. H. S. Waff and D. F. Weill, Electrical conductivity of magmatic liquids, effects of temperature, oxygen fugacity and composition, *Earth Planet. Sci. Lett.* **28**, 254–260 (1975).
30. H. Watanabe, Measurements of electrical conductivity of basalt at temperatures up to 1500°C and pressure to about 20 kilobars, *Spec. Contr. Geophys. Inst. Kyoto Univ.* **10**, 159–170 (1970).
31. N. T. Khitarov and A. V. Slutsky, Influence de la temperature et de la pression sur la conductibilite electrique de l'albite et du basalte, *J. Chim. Phys. et Phys. Chim.* **64**, 1085–1091 (1967).
32. I. Kushiro, Viscosity and structural changes of albite ($NaAlSi_3O_8$) melt at high pressures, *Earth Planet Sci. Lett.* **41**, 87–90 (1978).
33. I. Kushiro, Changes in viscosity and structure of melt of $NaAlSi_2O_6$ composition at high pressures, *J. Geophys. Res.* **81**, 6347–6350 (1976).
34. I. Kushiro, H. S. Yodder, and B. O. Mysen, Viscosities of basalt and andesite melts at high pressures, *J. Geophys. Res.* **81**, 6351–6356 (1976).
35. E. B. Lebedev and N. I. Khitarov, Influence of water on the electrical conductivity of silicate melts at high pressures, High Temperature High Pressure Electrochemistry in Aqueous Solutions N.A.C.E. at University of Surrey, England (1973).
36. A. T. Kuhn, ed., *Industrial Electrochemical Processes*, Elsevier, Amsterdam (1971).
37. C. E. Bowen, Production of H_2 and O_2 by electrolysis of H_2O, *Proc. Institution of Electrical Engineers* **90**, 474–485 (1943).
38. C. A. Angell, Electrical conductance of ionic liquids with water contents in the range 0–80 mol.%, *Aust. J. Chem.* **23**, 929–937 (1970).
39. D. A. Lown and H. R. Thirsk, Proton transfer conductance in aqueous solution, Parts 1 and 2, *Trans. Faraday Soc.* **67**, 132–152 (1971).
40. A. Reger, E. Peled, and E. Gileadi, Mechanism of high conductivity in a medium of low dielectric constant, *J. Phys. Chem.* **83**, 873–879 (1979).
41. A. Reger, E. Peled, and E. Gileadi, Determination of the nature of the ionic species in a low dielectric constant solvent from Transference number measurements, *J. Phys. Chem.* **83**, 869–873 (1979).
42. C. T. Moynihan, in *Ionic Interactions* (S. Petrucci, ed.), Vol. 1, Academic Press, New York (1971), Chapter 5.
43. C. T. Moynihan and R. W. Laity, Relative cation mobilities in potassium chloride–lithium chloride melts, *J. Phys. Chem.* **68**, 3312–3317 (1964).

44. E. R. Van Artsdalen and I. S. Yaffe, Electrical conductance and density of molten salt systems: KCl–LiCl, KCl–NaCl and KCl–KI, *J. Phys. Chem.* **59**, 118–127 (1955).

45. W. K. Behl and J. J. Egan, Transference numbers and ionic mobilities from electromotive force measurements on molten salt mixtures, *J. Phys. Chem.* **71**, 1764–1769 (1967).

46. H. H. Emons and H. Vogt, On the structure of charge-unsymmetrical salt melts of alkaline earth and alkali metal chlorides, *Z. Anorg. Allg. Chem.* **394**, 279–289 (1972).

47. D. S. Patterson and M. Chance, Production of sodium, British Patent No. 918,809 (1963).

48. Jacques van Diest, Process for the manufacture of sodium by electrolysis of fused salt bath, U.S. Patent No. 3,051,635 (1960).

49. A. V. Tomashov, V. A. Nichkov, A. E. Mordovin, and R. S. Khailikov, Interaction of potassium chlorides and beryllium chloride in melts and their mixtures, *Izv. Vyssh. Uchebn. Zaved. Tsvetn. Metall.* **5**, 81–85 (1975).

50. K. Grjotheim, C. Krohn, M. Malinovsky, K. Matiasovsky, and J. Thonstad, *Aluminium Electrolysis*, Aluminium-Verlag GmbH, Düsseldorf (1977).

51. E. W. Yim and M. Feinleib, Electrical conductivity of molten fluorides, *J. Electrochem. Soc.* **104**, 626–630 (1957).

52. K. Grjotheim, M. Malinovsky, and K. Matiasovsky, The effect of different additives on the conductivity of cryolite–alumina melts, *J. Metals* **21**, 28–33 (1969).

53. Chemistry Division, DSIR, Petone, New Zealand.

54. S. H. Wilson, Waiotapu Geothermal Field, New Zealand Department of Scientific and Industrial Research Bulletin 155 (1963a), pp. 87–118.

55. W. F. Giggenbach, The chemistry of Crater Lake, Mt. Ruapehu (New Zealand) during and after the 1971 active period, *N. Z. J. Sci.* **17**, 33–45 (1974).

Index